Forecasting Economic Time Series using Locally Stationary Processes

Volkswirtschaftliche Analysen

Herausgegeben von Elisabeth Allgoewer, Georg Hasenkamp, Wolfgang Maennig,
Christian Scheer und Peter Stahlecker

Band 19

PETER LANG
Frankfurt am Main · Berlin · Bern · Bruxelles · New York · Oxford · Wien

Tina Loll

Forecasting Economic Time Series using Locally Stationary Processes

A New Approach with Applications

PETER LANG
Internationaler Verlag der Wissenschaften

Bibliographic Information published by the Deutsche Nationalbibliothek
The Deutsche Nationalbibliothek lists this publication in the Deutsche Nationalbibliografie; detailed bibliographic data is available in the internet at http://dnb.d-nb.de.

Zugl.: Hamburg, Univ., Diss., 2011

D 18
ISSN 1432-8739
ISBN 978-3-631-62187-5

www.peterlang.de

to Hannes

Contents

1 Introduction

The assumption of stationarity plays a decisive role in time series analysis, and there exists a large amount of statistical techniques for stationary processes. In order to fit a certain model to a time series, i. e. to determine the model parameters reasonably, it is possible to use methods like *least squares* or *maximum likelihood* in the *time domain*, which means that the estimation is done using directly the observations. Furthermore, it is possible to compute estimations in the *frequency domain* using the spectral density of the respective process and fitting it to the periodogram, which can, for example, be done using the *Whittle likelihood*.

These techniques are very well investigated using asymptotic considerations, i. e. by examining the behaviour of the estimator if the length of the time series tends to infinity. In this way, one can show characteristics like asymptotic normality, unbiasedness, and efficiency. For general non-stationary processes however, contrary to the stationary case, future observations may not contain any information about the probabilistic structure of the process at present. That is why even in situations where it is obvious that a non-stationary model is more adequate, stationary models and techniques are used frequently (e. g. by looking at segments of the data). A more realistic approach is to assume that the second-order characteristics of a process vary over time.

In recent years there has been increasingly more awareness of the fact that data often show such time-varying second-order characteristics (see for example the findings of Pagan and Schwert 1990 for U.S. stock returns). As long as the period of the Great Depression is included in the series, their tests reject the hypothesis of covariance stationarity. Their results are confirmed by Loretan and Philips (1994) and by Van Bellegem and von Sachs (2002), who obtain similar results for exchange-rate data.

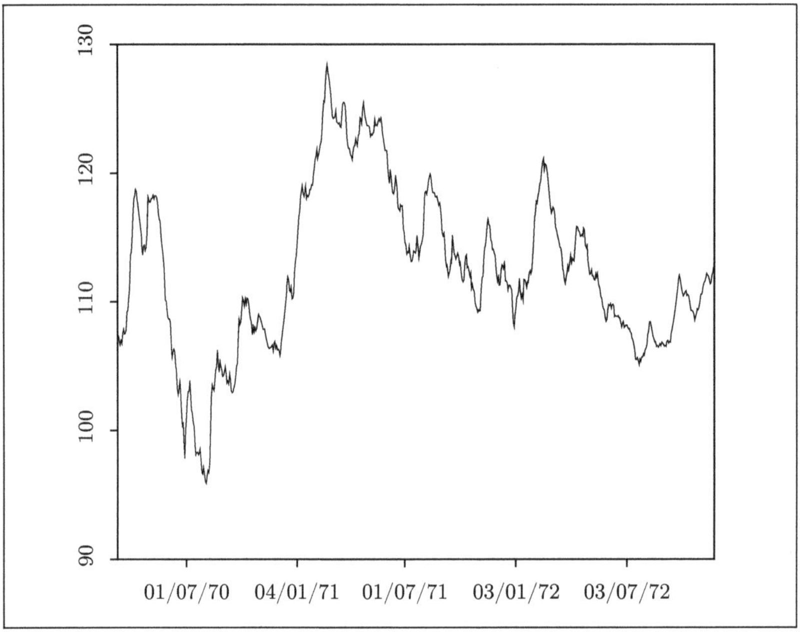

Figure 1.1: Daily index closing values for the Dow Jones Utility index between February 2nd 1970 and October 19th 1972

An example is given by the Dow Jones Utility index from February 2nd 1970 through October 19th 1972 in Figure 1.1. This index measures the performance of fifteen utility companies. It is obvious that this time series contains a trend, which can be removed by computing the first-order differences of the natural logarithm of the series (see Figure 1.2).

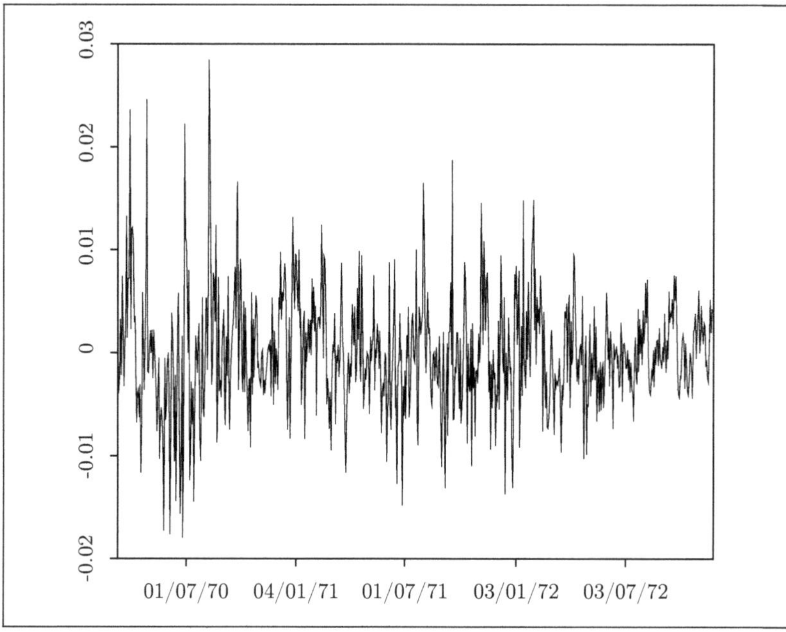

Figure 1.2: Log-returns of the Dow Jones index (calculated from the daily closing prices in Figure 1.1)

A commonly used model for stock indices is given by stationary autoregressive processes (AR). To illustrate the time dependence of the AR-coefficients and hence the non-stationarity of the covariances we estimate the coefficients of an AR(2) process for 50-day sections of the log-returns, i. e. we calculate the coefficients at times $t = 1, \dots, 50$, repeat this for $t = 2, \dots, 51$ and so forth. The smoothed results for the first coefficient can be found in Figure 1.3. In addition to the estimated first coefficient, the borders for 95% confidence intervals are given. The Figure shows that the coefficient is clearly non-zero. The hypothesis that it is constant over time cannot be rejected due to the relatively large values of the coefficient standard deviations. However, the estimated coefficient values obviously differ in time.

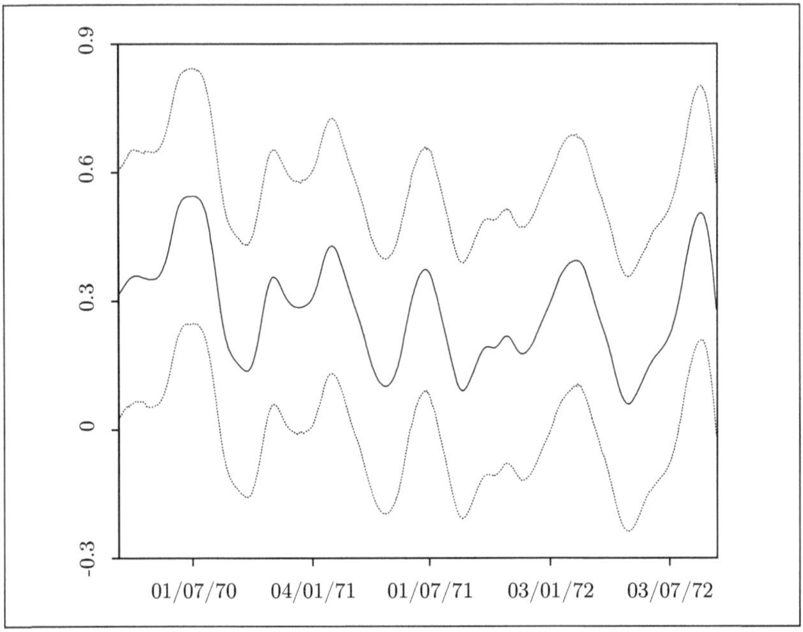

Figure 1.3: Smoothed coefficient estimations (parameter α_1) of AR(2) processes for the log-returns of the Dow Jones index

Priestley (1965) was the first one to consider stochastic processes that only locally display an approximately stationary behaviour. His concept of *oscillatory processes* with *evolutionary spectra* allows for the spectral analysis of certain non-stationary processes using a *time-dependent spectral representation* of the form

$$X_t = \int_{-\pi}^{\pi} \exp(i\lambda t)\, A_t(\lambda)\, \mathrm{d}Z(\lambda), \quad t \in \mathbb{Z} \tag{1.1}$$

with an orthogonal increment process $Z(\lambda)$ and a time-varying transfer function $A_t(\lambda)$. As it is not possible to consistently estimate a changing spectrum at a particular instant of time, it is necessary to calculate some kind of estimate of the *average* spectrum in the neighbourhood of this time instant. Therefore, it is assumed that the spectrum changes smoothly over time. However, the spectral representation from Equation (1.1) is not unique as it exists for different families $\mathcal{F} = \{\exp(i\lambda t)\, A_t(\lambda), t \in \mathbb{Z}\}$ of functions. Among these exists one, which oscillates least in t. According to Priestley, this family is approximated by statistical estimation as statistical procedures do have a smoothing effect in time. That is why he defines the class of *semi-stationary* processes as processes with the spectral representation from Equation (1.1), where A_t changes smoothly in time. However, within Priestley's concept asymptotic considerations are not possible, which is due to his approach to model non-stationarity. Nevertheless, it is necessary to construct an asymptotic theory for dealing with consistency and inference.

In order to solve this problem, Dahlhaus (1997) introduced the class of locally stationary processes. In his approach, the transfer function is rescaled in time and a doubly-indexed process is defined. Dahlhaus investigated statistical inference for locally stationary processes and showed that asymptotic results can be considered, if T is allowed to tend to infinity. Thereby, he uses an approach from the field of nonparametric regression. The time indices are rescaled, $u = \frac{1}{T}, \ldots, \frac{T}{T}$, i. e. allowing T to tend to infinity in this context does not mean looking into the future any more, but instead having more information about the local structure of the transfer function. The processes are doubly-indexed as they have a triangular form. For each element $\{X_{t,T}\}$, $t = 1, \ldots, T$, there exists a spectral representation

$$X_{t,T} = \int_{-\pi}^{\pi} \exp(i\lambda t)\, A_{t,T}^{\circ}(\lambda)\, \mathrm{d}Z(\lambda) \tag{1.2}$$

as with Priestley's approach. The processes from (1.2) form the foundation for Dahlhaus' definition of *locally stationary processes.* And there is a connection between the time-dependent spectral representations of the processes that are explained by the two approaches. (An exact definition will be given in Chapter 2.)

Our aim is to find out if it is possible to benefit from the local stationarity approach in the form of better forecasts for financial time series. As explained earlier, such time series are often modeled as AR(p) processes. A natural generalization of AR(p) processes, to deal with time-varying behaviour, are so-called TVAR(p) processes (*autoregressive processes with time-varying coefficients*), which are an important example of locally stationary processes. A semiparametric estimation procedure for fitting time-varying AR(p) processes to general non-stationary processes was introduced by Van Bellegem and Dahlhaus (2006). We want to investigate the forecasting abilities of this model using simulation studies and applying it to real financial data. As such a method does not exist so far, we focus on the development of a forecasting-procedure for TVAR processes. We thereby follow Dahlhaus' definition and assume that the true process is locally stationary. The models that we study are parameterized by a D-dimensional function $\boldsymbol{\theta}(u)$. The domains of the parameter functions $\boldsymbol{\theta}(u) = (\sigma, \alpha_1(u), \dots, \alpha_p(u))$ are rescaled to the interval $(0; 1]$. The models have the form

$$X_{t,T} + \sum_{j=1}^{p} \alpha_j\left(\frac{t}{T}\right) X_{t-j,T} = e_{t,T}, \quad t = 1, \dots, T,\, T > 0. \tag{1.3}$$

Here $e_{t,T}$ are independent, normally distributed random variables with $\mu = 0$ and a constant standard deviation σ.

This thesis is organized as follows:

The second chapter gives an introduction to the idea of locally stationary processes. First of all, the basic principles of stationary stochastic processes and their spectral representation are recalled. Then, the concept of evolutionary spectra, which forms the foundation for the approach of locally stationary processes, is introduced. This is followed by a formal definition of the class of locally stationary processes and an introduction of the class of time-varying autoregressive (*TVAR*) processes, an example from the class of locally stationary processes.

Chapter 3 presents a data-driven approach for fitting TVAR models to time series. This procedure was introduced by Van Bellegem and Dahlhaus (2006). It is constructed from a minimization of a penalized contrast function, which is an approximation to the Gaussian likelihood of the model. Thereby, the *method of sieves* is used to approximate the parameter curves, which we pay special intention to in the second part of the chapter.

In Chapter 4, after reviewing some fundamentals on predictions in the stationary case, we propose approaches for forecasting time series using autoregressive models with time-varying parameters. These are generalizations of the established Box-and-Jenkins-method. And, similar to this method, we recommend an iterative approach to fit the models that are used for computing the predictions. Finally, we evaluate the forecasts from the proposed procedures by means of Monte Carlo simulations.

Chapter 5 describes the practical application of TVAR processes to financial time series, namely the log-returns of the Euro-Bund-Future and the log-returns of the Dow Jones Utility index, since in recent years it has been observed that these data often show time-varying second-order characteristics.

Concluding remarks are given in Chapter 6 summarizing the most important contributions of this work and identifying remaining problems for future research.

2

From stationarity to local stationarity

In this chapter we give an introduction to the idea of locally stationary processes. The course of action is the following: After reviewing the fundamentals of stationary stochastic processes and their spectral representation, we recall the concept of evolutionary spectra. This concept gives the basic principles for the approach of locally stationary processes. It is followed by a formal definition of the class of locally stationary processes. As we will see, this definition is motivated by the relation between stationarity and the existence of a spectral representation. Finally, we introduce the class of time-varying autoregressive processes, which is an important example from the class of locally stationary processes.

2.1 Stationary stochastic processes

2.1.1 A short introduction to stationarity

This work is concerned with data coming in the form of a set of observations x_t, each one recorded at a specified time t. A sequence of these observations x_t is called *time series*. Thereby, the set $\mathcal{T}$ of time indices obeys the relation $\mathcal{T} \subseteq \mathbb{R}$.

An important step in the analysis of time series is the selection of a suitable model for the data. A natural and general ansatz is to suppose that every observation x_t is a realized value of a certain random variable X_t. The time series $\{x_t : t \in \mathcal{T}\}$ is then a realization of the sequence of random variables $\{X_t : t \in \mathcal{T}\}$. These considerations suggest modelling the data as a realization of a *stochastic process* $\{X_t : t \in \mathbb{R}\}$. To clarify this idea we need to define what is meant by a stochastic process.

Definition 1

Given the probability space $(\Omega, \mathcal{A}, P)$, *a real* stochastic process *is a sequence* $\{X_t : t \in \mathcal{T}\}$ *of functions* $X_t := X(t,\omega)$ *defined on* $\mathcal{T} \times \Omega$ *such that for fixed* t, $X(t,\omega)$ *is a random variable on* $(\Omega, \mathcal{A}, P)$.

For fixed ω, $X(t,\omega)$ is a function of t called *realization.* Under the condition that the observations x_t have been generated by a stochastic process $\{X_t\}$, the time series $\{x_t\}$ is a realization of this stochastic process. For convenience we will write $\{x_t\}$ instead of $\{x_t : t \in \mathcal{T}\}$ and $\{X_t\}$ instead of $\{X_t : t \in \mathcal{T}\}$ respectively.

As we will see in the next section, the definition of the spectral representation of a stationary stochastic process, which plays a key role in the definition of the class of locally stationary processes, requires the definition of complex stochastic processes.

Definition 2

A complex stochastic process *is defined by*

$$Y_t = U_t + iV_t, \quad t \in \mathcal{T}$$

where $\{U_t\}$ *and* $\{V_t\}$ *are real stochastic processes. Then the mean function and the autocovariance function are defined by*

$$\mu_t = \mathrm{E}(Y_t) = \mathrm{E}(U_t) + i\mathrm{E}(V_t) \tag{2.1}$$

and

$$\gamma(t, t+k) = \mathrm{E}[(Y_t - \mu_t)\overline{(Y_{t+k} - \mu_{t+k})}]. \tag{2.2}$$

The overbar denotes the operation of conjugation, and it is $i^2 = -1$.

If further $\mathrm{E}(|Y_t|^2) < \infty$ holds for every $t \in \mathbb{R}$, $\{Y_t\}$ is called L^2 *process.* The term L^2 *process* emphasizes that the process is defined on an L^2 space. In what follows $L^2_{\mathbb{C}}$ denotes the L^2 space of square integrable and complex valued random variables.

Definition 3
Let $X, Y \in L^2_{\mathbb{C}}$, where X and Y are random variables. Then

$$\langle X, Y\rangle := \mathrm{E}(X\overline{Y}) \tag{2.3}$$

is an inner product on $L^2_{\mathbb{C}}$. Further

$$||Y||_2 := \langle Y, Y\rangle^{\frac{1}{2}}$$

is defined for every $Y \in L^2_{\mathbb{C}}$. Then $L^2_{\mathbb{C}}$ is a Hilbert space.

In what follows we will assume (if nothing else is remarked) that the random variables used form complex valued Hilbert spaces. In the definition of stationarity for these processes, we distinguish between *strict* stationarity and *weak* stationarity.

Definition 4
A stochastic process is called strictly stationary, *if*

$$\begin{aligned} &G_{Y_{t_1},Y_{t_2},\ldots,Y_{t_T}}(y_{t_1}, y_{t_2}, \ldots, y_{t_T}) \\ &= G_{Y_{t_1+\tau},Y_{t_2+\tau},\ldots,Y_{t_T+\tau}}(y_{t_1}, y_{t_2}, \ldots, y_{t_T}) \end{aligned}$$

for all possible sets of indices $t_1, t_2, \ldots, t_T$ and $t_1+\tau, t_2+\tau, \ldots, t_T+\tau$ in the index set and all $y_{t_1}, y_{t_2}, \ldots, y_{t_T}$ in the range of the random variable Y_t.

Thereby,

$$\begin{aligned} &G_{Y_{t_1},Y_{t_2},\ldots,Y_{t_T}}(y_{t_1}, y_{t_2}, \ldots, y_{t_T}) \\ &:= \mathrm{P}\{\omega : Y(t_1,\omega) \le y_{t_1}, Y(t_2,\omega) \le y_{t_2}, \ldots, Y(t_T,\omega) \le y_{t_T}\} \end{aligned}$$

is the *joint distribution function* of $\{Y_{t_1}, Y_{t_2}, \ldots, Y_{t_T}\}$ for a set of random variables $\{Y_{t_1}, Y_{t_2}, \ldots, Y_{t_T}\}$.

If the second-order moments exist, strict stationarity implies that

$$\mu_t = \mu_{t+\tau} = \mu \tag{2.8}$$

and

$$\gamma(t, t+k) = \gamma(t+\tau, t+k+\tau) = \gamma(k) \tag{2.9}$$

for all $t, \tau \in \mathbb{R}$. These properties suffice for many considerations, which leads to the following definition.

Definition 5
A stochastic process is called weakly stationary *if the mean function and the covariance function as defined in* (2.1) *and* (2.2) *exist and satisfy* (2.8) *and* (2.9).

In what follows we will only regard weakly stationary processes, which will be called *stationary* processes from now on.

2.1.2 Spectral representation of stationary processes

Below we give a definition for the *spectral representation* of stationary processes. To derive such a definition, the following theorems are required.

Theorem 1
A sequence of autocovariances $\gamma(k)$ of a stationary process is positive definite, i. e. $\gamma(k)$ is a real function with $\gamma(k) = \gamma(-k)$ and

$$\sum_{j,k=-\infty}^{\infty} \gamma(k-j)\alpha_j \overline{\alpha_k} \geq 0$$

for every absolute summable complex sequence $\{\alpha_j\}$ (see e. g. Dhrymes 1974, p. 399 for the proof).

Theorem 2
If $\gamma(k)$ is a positive definite sequence, there exists a stationary process $\{Z_t\}$, $t \in \mathbb{N}$, such that $\gamma(k)$ is its autocovariance function (see again Dhrymes 1974, p. 399).

Theorem 3
A function $\gamma(k)$ on $\mathbb{N}$ is positive definite if and only if there exists a function F so that

$$\gamma(k) = \int_{-\pi}^{\pi} \exp(ik\lambda)\,\mathrm{d}F(\lambda)$$

holds for a suitable function, where F defined for $|\lambda| \leq \pi$ is monotone nondecreasing (see e. g. Doob 1953, pp. 473 et seqq. for the proof).

Definition 6
The function $F(\cdot)$ *is called* spectral distribution function *and, if it exists, its derivative* $f(\cdot) = F'(\cdot)$ *is called* spectral density function.

The next two theorems are given to illustrate the relationship between stationarity and the existence of a spectral representation. An extension of this relationship will be given in Section 2.2.2.

Theorem 4
If $\{X_t\}$ *is a stationary process with* $t \in \mathbb{Z}$ *and zero mean function, there exists a process* $\{Z(\lambda) : -\pi < \lambda \leq \pi\}$ *such that*

$$X_t = \int_{-\pi}^{\pi} \exp(it\lambda)\, \mathrm{d}Z(\lambda), \tag{2.10}$$

where $Z(\lambda)$ *is a stochastic process with orthogonal increments, i. e. for* $-\pi < \phi \leq \pi$

$$\mathrm{E}[|Z(\lambda) - Z(\phi)|] < \infty$$

and if whenever $\phi_1 < \lambda_1 \leq \phi_2 < \lambda_2$

$$\mathrm{E}[(Z(\lambda_2) - Z(\phi_2))\overline{(Z(\lambda_1) - Z(\phi_1))}] = 0.$$

The process Z *has the properties*

$$\mathrm{E}[\mathrm{d}Z(\lambda)\overline{\mathrm{d}Z(\lambda)}] = \mathrm{d}F(\lambda)$$

and

$$\mathrm{E}[\mathrm{d}Z(\lambda_1)\overline{\mathrm{d}Z(\lambda_2)}] = 0 \quad \text{for} \quad \lambda_1 \neq \lambda_2,$$

where $F(\cdot)$ *is the spectral distribution of* $\{X_t\}$*. If* $\{X_t\}$ *is real, then there exist real stochastic processes*

$$\{u(\lambda) : 0 < \lambda \leq \pi\},\ \{v(\lambda) : 0 < \lambda \leq \pi\}$$

with orthogonal increments such that

$$X_t = \int_0^{\pi} \cos(t\lambda)\, \mathrm{d}u(\lambda) + \int_0^{\pi} \sin(t\lambda)\, \mathrm{d}v(\lambda).$$

For the processes u and v it is

$$\mathrm{E}[\mathrm{d}u(\lambda)]^2 = \mathrm{E}[\mathrm{d}v(\lambda)]^2 = \mathrm{d}F(\lambda) \quad \textit{for} \quad 0 < \lambda \leq \pi$$

and

$$\mathrm{E}[\mathrm{d}u(\lambda_1)\mathrm{d}u(\lambda_2)] = \mathrm{E}[\mathrm{d}v(\lambda_1)\mathrm{d}v(\lambda_2)] = \mathrm{E}[\mathrm{d}u(\lambda)\mathrm{d}v(\lambda)] = 0$$

for $\lambda_1 \neq \lambda_2$, *where* $F(\cdot)$ *is the spectral distribution of* $\{X_t\}$.

(For a proof see Doob 1953, pp. 481–488.)

The above theorem shows that in the case of stationarity X_t can be written as a stochastic integral of the function $f(\lambda) = \exp(it\lambda)$ with respect to the process $Z(\lambda)$. A detailed explanation of what is meant by a *stochastic process* and its connection to stochastic differential equations is given by Iacus (2008, Section 1.9). We will outline only the most important points. The stochastic integral of the function $f(\lambda)$ is defined as the limit to the sequence given through the stochastic integrals of the sequences $f^{(n)}(\lambda)$ that are defined by $f^{(n)}(\lambda) := f(\lambda_j)$, $\lambda_j \leq \lambda < \lambda_{j+1}$. Thereby, the stochastic integral of $f^{(n)}$ is defined as

$$I(f^{(n)}) := \sum_{j=0}^{n-1} f^{(n)}(\lambda_j)\{Z(\lambda_{j+1}) - Z(\lambda_j)\} = \sum_{j=0}^{n-1} f(\lambda_j)\{Z(\lambda_{j+1}) - Z(\lambda_j)\}.$$

It can be shown that $I(f^{(n)})$ converges in quadratic mean to (2.10).

Theorem 5
Let

$$X_t = \int_{-\pi}^{\pi} \exp(it\lambda)\,\mathrm{d}Z(\lambda), \tag{2.11}$$

where $\{Z(\lambda)\}$ *has orthogonal increments with*

$$\mathrm{E}[\mathrm{d}Z(\lambda)\,\overline{\mathrm{d}Z(\lambda)}] = \mathrm{d}F(\lambda).$$

Then the autocovariance function

$$\gamma(k) = \int_{-\pi}^{\pi} \exp(ik\lambda)\,\mathrm{d}F(\lambda)$$

is independent of t, so that the process $\{X_t\}$ *is stationary with spectral distribution function* F.

(See for example Dhrymes 1974, pp. 404–407.)

We obtain the autocovariance function $\gamma(k)$ of the process $\{X_t\}$ with zero mean function by computing

$$\gamma(k) = \mathrm{E}(X_{t+k}\overline{X_t})$$

as defined in (2.2). As the random variables of the process $\{X_t\}$ are defined on $L_{\mathbb{C}}$, we obtain $\gamma(k)$ with (2.3) by computing the inner product of X_{t+k} and X_t. With (2.11) we derive

$$\langle X_{t+k}, X_t\rangle = \left\langle \int_{-\pi}^{\pi} \exp(i(t+k)\lambda)\, \mathrm{d}Z(\lambda), \int_{-\pi}^{\pi} \exp(it\lambda)\, \mathrm{d}Z(\lambda) \right\rangle .$$

From the rules of calculation for stochastic integrals (cf. Kreiß and Neuhaus 2006, p. 82) it follows that $\gamma(k)$ has the form

$$\gamma(k) = \int_{-\pi}^{\pi} \exp(ik\lambda)\, \mathrm{d}F(\lambda).$$

So for every stationary process $\{X_t\}$ with zero mean function there exists a process with orthogonal increments $\{Z(\lambda)\}$ with $-\pi < \lambda \leq \pi$, such that

$$X_t = \int_{-\pi}^{\pi} \exp(it\lambda)\, \mathrm{d}Z(\lambda), \tag{2.12}$$

and moreover any process $\{X_t\}$ defined by Equation (2.12) with the stated conditions on $\{Z(\lambda)\}$ is stationary.

From now on we will only regard processes with zero mean function.

2.1.3 Stationary ARMA processes

An important class of stationary processes is given by *stationary autoregressive moving average (ARMA) processes*. For any autocovariance function $\gamma(\cdot)$ such that $\lim_{h\to\infty} \gamma(h) = 0$, and for any integer $k > 0$, it is possible to find an ARMA process with autocovariance function $\gamma_X(\cdot)$, such that $\gamma_X(h) = \gamma(h),\ h = 0, 1, \ldots, k$. This is one reason for which the family of ARMA processes plays a key role in the modelling of time series data. Besides, the linear structure of ARMA processes leads to a very simple theory of linear prediction.

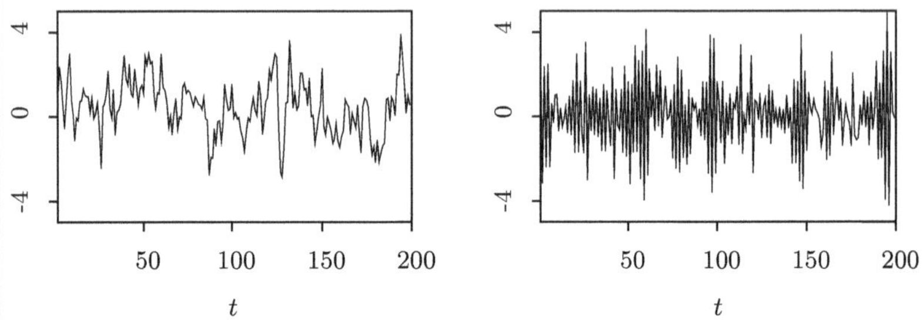

(a) $\alpha_1 = 0.8$ (b) $\alpha_1 = -0.8$

Figure 2.1: Two hundred observations from a realization of the process $X_t = \alpha_1 X_{t-1} + e_t$

White noise processes

The simpliest kind of process is one for which the random variables are independently and identically distributed with zero mean and constant variance.

Definition 7
A sequence of uncorrelated random variables with mean zero and finite variance, $\sigma^2 > 0$, is called white noise.

Since this stochastic process is of particular importance, as a very wide class of stationary processes can be generated by using white noise processes, we shall reserve the symbol $\{e_t\}$ for it.

Autoregressive processes

The current value of an autoregressive process is expressed as a weighted sum of past values plus the current shock. Within an autoregressive process of order p, X_t can be considered to be regressed on the p previous X_t's. That is why the process is called *autoregressive*.

Definition 8
A stochastic process $\{X_t\}_{t\in\mathbb{Z}}$*, for which*

$$\sum_{i=0}^{p} \alpha_i X_{t-i} = e_t \tag{2.13}$$

holds with $\alpha_0 \neq 0$*,* $\alpha_p \neq 0$*, and* $\{e_t\}_{t\in\mathbb{Z}}$ *is a white noise process, is called* pth order autoregressive process.

See Figure 2.1 for two realizations of AR(1) processes.

Moving average processes

Definition 9
If M *is a positive integer,* $\beta_j \in \mathbb{R}$*,* $\beta_0 \neq 0$*,* $\beta_M \neq 0$*, and* $\{e_t\}$ *is a white noise process, the stochastic process* $\{X_t\}$ *defined by*

$$X_t = \beta(L)e_t, \tag{2.14}$$

where $\beta(L)$ *is a linear filter defined by*

$$\beta(L)e_t := \sum_{j=0}^{M} \beta_j e_{t-j},$$

is called a finite moving average process *of order* M*. Thereby, the operator* L *is the so called* lag operator*, that replaces* X_t *by* X_{t-1}*, i. e.*

$$LX_t := X_{t-1}.$$

Against it, if there exists no M*, such that* $\beta_j = 0$ *for all* j *for which* $|j| > M$*,* $\{X_t\}$ *is called* infinitive moving average process *and has the representation*

$$X_t = \sum_{j=0}^{\infty} \beta_j e_{t-j}.$$

Two realizations of MA(1) processes can be found in Figure 2.2.

The MA process is said to be *invertible* if Equation (2.14) can also be written as

$$\beta^{-1}(L)X_t = e_t. \tag{2.15}$$

Here, $\beta^{-1}(L)$ is the inverse of the linear filter $\beta(L)$. This is the case if and only if the zeros of $\beta(L)$ lie outside the unit circle. Equation (2.15) shows that an MA(q) process is equivalent to an AR process of infinite order. In the same way, a stationary AR(p) process can be written as an MA(∞) process. Thus, autoregressive processes and moving average processes are to some extent equivalent.

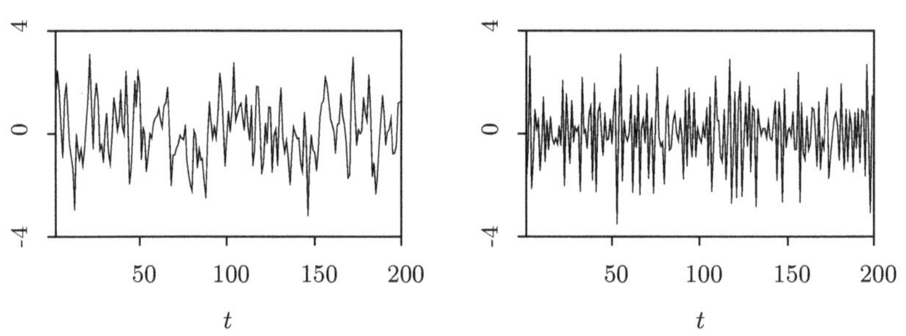

(a) $\beta_1 = 0.8$ **(b)** $\beta_1 = -0.8$

Figure 2.2: Two hundred observations from a realization of the process $X_t = \beta_1 e_{t-1} + e_t$

A reasonable extension to the models from Definitions 8 and 9 is given by the following definition.

Definition 10
A mixed model of the form

$$X_t = \alpha_1 X_{t-1} + \ldots + \alpha_p X_{t-p} + e_t + \beta_1 e_{t-1} + \ldots + \beta_q e_{t-q}$$

is called ARMA(p, q) model.

The next theorem provides the spectrum of such an ARMA(p, q) process.

Theorem 6
Let $\{X_t\}$ be an ARMA process as defined in Definition 10. Then the theoretical spectrum of the process as defined in Definition 6 has the form

$$f(\lambda) = \frac{|1 - \sum_{u=1}^{q} \beta_u \exp(i2\pi\lambda u)|^2}{|1 - \sum_{\nu=1}^{p} \alpha_\nu \exp(i2\pi\lambda\nu)|^2}\sigma^2. \tag{2.16}$$

2.1.4 Asymptotical properties of the sample partial autocorrelations of a stationary AR(p) process

The *partial autocorrelation function*, like the autocorrelation function, depends only on the second order properties of a process. It conveys important information concerning the dependence structure of a stationary process. The partial correlation at lag k may be understood as the correlation between X_1 and X_{k+1} adjusted for the observations $X_2, \ldots, X_k$. This is specified in the following definition.

Definition 11
Let $\{X_t\}_{t\in\mathbb{Z}}$ be a stationary process. Then

$$\phi(k) := \mathrm{Cor}(X_t - \hat{X}_t, X_{t-k} - \hat{X}_{t-k})$$

is called partial autocorrelation, *where $\hat{X}_t$ and $\hat{X}_{t-k}$ are the best linear approximations of X_t and X_{t-k} by the random variables $X_{t-k+1}, \ldots, X_{t-1}$ between the time points $t-k$ and t, $k > 1$, $\phi(0) := 1$ and $\phi(1) := \rho(1)$. Here $\hat{X}_t = \alpha_0 + \alpha_1 X_{t-1} + \alpha_2 X_{t-2} + \cdots + \alpha_{k-1} X_{t-k+1}$ is called* best linear approximation *of X_t by $X_{t-k+1}, \ldots, X_{t-1}$ if*

$$\mathrm{E}(X_t - \alpha_0 - \alpha_1 X_{t-1} - \cdots - \alpha_{k-1} X_{t-k+1})^2$$

is minimized over $\alpha_0, \alpha_1, \ldots, \alpha_{k-1}$.

We will now give an equivalent definition of the partial correlation function. (For a proof of the equivalence of the two definitions see for example Brockwell and Davis 1991, p. 171.)

If the *autocorrelation matrix* for a stationary time series of length T is given by

$$\boldsymbol{P}_T = \begin{pmatrix} 1 & \gamma(1) & \gamma(2) & \dots \gamma(T-1) \\ \gamma(1) & 1 & \gamma(1) & \dots \gamma(T-2) \\ \gamma(2) & \gamma(1) & 1 & \dots \gamma(T-3) \\ \vdots & \vdots & & \ddots \quad \vdots \\ \gamma(T-1) & \gamma(T-2) & \gamma(T-3) & \dots \quad 1 \end{pmatrix},$$

the *partial autocorrelation* $\phi(k)$ is defined as

$$\phi(k) = \frac{\det \boldsymbol{P}_k^*}{\det \boldsymbol{P}_k},$$

where $\boldsymbol{P}_k$ is the $k \times k$ autocorrelation matrix, and $\boldsymbol{P}_k^*$ is $\boldsymbol{P}_k$ with the last column replaced by

$$\begin{pmatrix} \gamma(1) \\ \vdots \\ \gamma(k) \end{pmatrix}.$$

Theorem 7
$\{X_t\}$ is an AR(p) process if and only if $\phi(p) \neq 0$ and $\phi(\tau) = 0$ for $\tau > p$.

Theorem 8
Let $\alpha(L)X_t = e_t$ be a stationary AR(p) process with $\mathrm{E}(e_t) = 0$ and $\mathrm{Var}(e_t) = \sigma^2 < \infty$. Then the sample partial autocorrelations $\hat{\phi}(\tau)$ of order $\tau > p$ for a time series of length T, which is (part of) a realization of that process, are asymptotically independent normally distributed with mean 0 and variance $1/T$.

We discuss the proof in-depth as we want to generalize it to the case of locally stationary AR processes later. (The proceeding is taken from Hannan 1970, pp. 329 et seqq.)

Proof. An AR(p) process can also be written as AR(τ) process with $\alpha_{p+1} = \alpha_{p+2} = \ldots = \alpha_\tau = 0$. Then, the estimated partial autocorrelation $\hat{\phi}(\tau)$ equates to the Yule-Walker estimator[1] $\hat{\alpha}_\tau$ of α_τ. Let

[1] See for example Schlittgen and Streitberg (2001, pp. 253 et seqq.) for an introduction to the theory of the Yule-Walker estimator.

$\boldsymbol{\Sigma} = (\sigma_{uv})$ be the covariance matrix of τ succeeding observations of the process and $\boldsymbol{\Sigma}^{-1} = (\sigma^{uv})$ its inverse.

Equation (2.13) can be regarded as a regression model. This suggests to estimate the parameters following the least squares approach which leads to the *normal equations*

$$\begin{aligned} c_{11}\hat{\alpha}_1 + \ldots + c_{1p}\hat{\alpha}_p &= c_{01} \\ c_{21}\hat{\alpha}_1 + \ldots + c_{2p}\hat{\alpha}_p &= c_{02} \\ &\vdots \\ c_{p1}\hat{\alpha}_1 + \ldots + c_{pp}\hat{\alpha}_p &= c_{0p} \end{aligned} \qquad (2.17)$$

with

$$c_{ij} = \sum_{t=p+1}^{T} x_{t-i}x_{t-j}.$$

A slightly modified version of (2.17) is given by

$$\sum_{j=1}^{p} \tilde{c}_{ij}\hat{\alpha}_j = \tilde{c}_{i0}, \quad i = 1, \ldots, p \qquad (2.18)$$

with

$$\tilde{c}_{ij} := \frac{1}{T}\sum_{t=1}^{T} X_{t-i}X_{t-j}.$$

This modification is asymptotically irrelevant.

Since we have

$$\sum_{j=1}^{p} X_{t-j}\alpha_j = X_t - e_t,$$

it is

$$\sum_{j=1}^{p} \tilde{c}_{ij}\alpha_j = \frac{1}{T}\sum_{t=1}^{T} X_{t-i}\sum_{j=1}^{p} X_{t-j}\alpha_j$$
$$= \frac{1}{T}\sum_{t=1}^{T} X_{t-i}X_t - \frac{1}{T}\sum_{t=1}^{T} X_{t-i}e_t$$
$$= \tilde{c}_{i0} - \frac{1}{\sqrt{T}}h_i$$

with

$$h_i := \frac{1}{\sqrt{T}}\sum_{t=1}^{T} X_{t-i}e_t, \quad i = 1, 2, \ldots, p.$$

Using (2.18) we get

$$\sqrt{T}\sum_{j=1}^{p} \tilde{c}_{ij}(\hat{\alpha}_j - \alpha_j) = \sqrt{T}\tilde{c}_{i0} - \sqrt{T}\sum_{j=1}^{p} \tilde{c}_{ij}\alpha_j$$
$$= \sqrt{T}\tilde{c}_{i0} - \sqrt{T}\tilde{c}_{i0} + h_i$$
$$= h_i.$$

In matrix notation, with $\boldsymbol{h} = (h_1, \ldots, h_p)'$ and $\hat{\boldsymbol{\Sigma}}_p = (\tilde{c}_{ij})$, this corresponds to

$$\hat{\boldsymbol{\Sigma}}_p\sqrt{T}(\hat{\boldsymbol{\alpha}} - \boldsymbol{\alpha}) = \boldsymbol{h}.$$

The substitution of e_t by $X_t - \alpha_1 X_{t-1}$ in the definition of h_i reveals that h_i is a linear function of the $\sqrt{T}\tilde{c}_{ij}$. If these are multivariate normally distributed, $\boldsymbol{h}$ is also multivariate normally distributed.

Obviously, $\mathrm{E}(\boldsymbol{h}) = \mathbf{0}$ holds and the covariances have the form

$$\mathrm{E}(h_i h_j) = \frac{1}{T}\sum_{s=1}^{T}\sum_{t=1}^{T} \mathrm{E}(e_s e_t X_{s-i} X_{t-j})$$
$$= \frac{1}{T}\sum_{s=1}^{T} \sigma^2 c_{|i-j|} = \sigma^2 c_{|i-j|}.$$

Consequently, the covariance matrix $\mathrm{Var}(\boldsymbol{h})$ of $\boldsymbol{h}$ equates to $\sigma^2\boldsymbol{\Sigma}_p$. $\hat{\boldsymbol{\Sigma}}_p$ converges with probability 1 to $\boldsymbol{\Sigma}_p$. Hence, $T^{1/2}(\hat{\boldsymbol{\alpha}} - \boldsymbol{\alpha}) = \hat{\boldsymbol{\Sigma}}_p^{-1}\boldsymbol{h}$ follows the same distribution as $\boldsymbol{\Sigma}_p^{-1}\boldsymbol{h}$, which is according to the above considerations a multivariate normal distribution with covariance matrix $\boldsymbol{\Sigma}_p^{-1}\mathrm{Var}(\boldsymbol{h})\boldsymbol{\Sigma}_p^{-1} = \sigma^2\boldsymbol{\Sigma}_p^{-1}$.

Thus, it is asymptotically

$$\hat{\phi}(\tau) \quad \dot{\sim} \quad \mathcal{N}(\phi(\tau), \frac{\sigma^2\sigma^{\tau\tau}}{\sqrt{T}}).$$

The covariance matrix of a vector $\boldsymbol{x} = (X_1, \ldots, X_n)'$ of random variables is

$$\mathrm{Var}(\boldsymbol{x}) = \boldsymbol{\Sigma}_{\boldsymbol{xx}} = \begin{pmatrix} \mathrm{Cov}(X_1, X_1), \ldots, \mathrm{Cov}(X_1, X_n) \\ \vdots \quad\quad \ddots \quad\quad \vdots \\ \mathrm{Cov}(X_n, X_1), \ldots, \mathrm{Cov}(X_n, X_n) \end{pmatrix}.$$

Whereas for $\boldsymbol{x}$ and a single random variable Y the covariance matrix is

$$\mathrm{Cov}(\boldsymbol{x}, Y) = \boldsymbol{\Sigma}_{\boldsymbol{xy}} = (\mathrm{Cov}(X_1, Y), \ldots, \mathrm{Cov}(X_n, Y))' = \boldsymbol{\Sigma}'_{\boldsymbol{yx}}.$$

If $\boldsymbol{\Sigma}_{\boldsymbol{xx}}$ and $\boldsymbol{\Sigma}_{\boldsymbol{yy.x}} = \boldsymbol{\Sigma}_{\boldsymbol{yy}} - \boldsymbol{\Sigma}_{\boldsymbol{yx}}\boldsymbol{\Sigma}_{\boldsymbol{xx}}^{-1}\boldsymbol{\Sigma}_{\boldsymbol{xy}}$ are nonsingular, we have

$$\begin{pmatrix} \boldsymbol{\Sigma}_{\boldsymbol{yy}} & \boldsymbol{\Sigma}_{\boldsymbol{yx}} \\ \boldsymbol{\Sigma}_{\boldsymbol{xy}} & \boldsymbol{\Sigma}_{\boldsymbol{xx}} \end{pmatrix}^{-1} = \begin{pmatrix} \boldsymbol{\Sigma}_{\boldsymbol{yy.x}}^{-1} & -\boldsymbol{\Sigma}_{\boldsymbol{yy.x}}^{-1}\boldsymbol{\Sigma}_{\boldsymbol{yx}}\boldsymbol{\Sigma}_{\boldsymbol{xx}}^{-1} \\ -\boldsymbol{\Sigma}_{\boldsymbol{xx}}^{-1}\boldsymbol{\Sigma}_{\boldsymbol{xy}}\boldsymbol{\Sigma}_{\boldsymbol{yy.x}}^{-1} & \boldsymbol{\Sigma}_{\boldsymbol{xx}}^{-1} + \boldsymbol{\Sigma}_{\boldsymbol{xx}}^{-1}\boldsymbol{\Sigma}_{\boldsymbol{xy}}\boldsymbol{\Sigma}_{\boldsymbol{yy.x}}^{-1}\boldsymbol{\Sigma}_{\boldsymbol{yx}}\boldsymbol{\Sigma}_{\boldsymbol{xx}}^{-1} \end{pmatrix}. \tag{2.19}$$

Thereby, $\boldsymbol{\Sigma}_{\boldsymbol{yy.x}}$ is the *partial covariance matrix of* $\boldsymbol{y}$ *after eliminating* $\boldsymbol{x}$, i. e. the covariance matrix $\mathrm{Var}(\boldsymbol{y} - \hat{\boldsymbol{y}})$ of $\boldsymbol{y} - \hat{\boldsymbol{y}}$, where $\hat{\boldsymbol{y}}$ is the best linear approximation of $\boldsymbol{y}$ by $\boldsymbol{x}$. (For the proof see Schlittgen and Streitberg 2001, p. 526.)

Setting $Y = (X_\tau)$ and $\boldsymbol{x} = (X_1, \ldots, X_{\tau-1})'$, we get from (2.19)

$$\sigma^{\tau\tau} = \boldsymbol{\Sigma}_{\boldsymbol{yy.x}}^{-1}.$$

As in our case $\boldsymbol{y} = Y = X_\tau$, $\boldsymbol{\Sigma}_{\boldsymbol{yy.x}}$ corresponds to the partial variance $\sigma_{\tau\tau.x}$.

If we define the *coefficient of determination* $R^2_{i.x}$ as

$$R^2_{i.x} := \frac{\sigma_{ii} - \sigma_{ii.x}}{\sigma_{ii}},$$

we get

$$\sigma^{ii} = \sigma^{-1}_{ii.x} = \sigma^{-1}_{ii} \frac{1}{1 - R^2_{i.x}}.$$

Since the coefficient of determination equates to the squared multiple correlation coefficient, for the main diagonal elements of the inverse of the covariance matrix $\boldsymbol{\Sigma}^{-1} = (\sigma^{ij})$ of the random variables $X_1, X_2, \ldots, X_n$ it is

$$\sigma^{ii} = \sigma^{-1}_{ii}(1 - \tilde{R}^2_i)^{-1},$$

in which $\tilde{R}_i$ is the multiple correlation of a variable X_i with the other $n - 1$ variables. Therefore, we get

$$\begin{aligned} \sigma^{\tau\tau} &= \frac{1}{\sigma_{\tau\tau}(1 - R^2_{\tau.12\ldots\tau-1})} \\ &= \frac{1}{\gamma(0)\, Q(\tau - 1)}. \end{aligned}$$

Thereby,

$$Q(\tau - 1) := \frac{\mathrm{Var}(X_\tau - \hat{X}_{\tau-1}(1))}{\mathrm{Var}(X_\tau)},$$

where $\mathrm{Var}(X_\tau - \hat{X}_{\tau-1}(1))$ is the error of prediction.

The best linear approximation of X_τ by $X_1, X_2, \ldots, X_{\tau-1}$ is

$$X_\tau - \hat{X}_\tau = X_\tau - \alpha_1 X_{\tau-1} - \alpha_2 X_{\tau-2} + \ldots + \alpha_p X_{\tau-p} = e_\tau.$$

From this follows

$$\gamma(0) \cdot Q(\tau - 1) = \mathrm{Var}(X_\tau - \hat{X}_\tau) = \mathrm{Var}(\epsilon_\tau) = \sigma^2.$$

□

2.2 Locally stationary processes

2.2.1 Evolutionary spectrum

If the stationary assumption is withdrawn Definition 6 is not applicable since there is no autocovariance function any more. In order to extend the spectral representation to non-stationary processes, Priestley introduced the concept of *evolutionary spectra.* His approach allows for the spectral analysis of non-stationary processes using a *time-dependent spectrum* (see Priestley 1965). He generalized the spectral decomposition of a stationary series in (2.10) by introducing a time-varying transfer function $A_t(\lambda)$ with the help of which we get

$$X_t = \int_{-\pi}^{\pi} \exp(i\lambda t)\, A_t(\lambda)\, \mathrm{d}Z(\lambda), \quad t \in \mathbb{Z}, \tag{2.20}$$

where $Z(\lambda)$ has again orthogonal increments and it is

$$\mathrm{E}[\mathrm{d}Z(\lambda)\, \overline{\mathrm{d}Z(\lambda)}] = f(\lambda)\, \mathrm{d}\lambda.$$

Here $f(\lambda)$ is the spectrum of the stationary process

$$\int_{-\pi}^{\pi} \exp(it\lambda)\, \mathrm{d}Z(\lambda).$$

The evolutionary spectral density is defined as

$$f_t(\lambda) = |A_t(\lambda)|\, f(\lambda).$$

Obviously, the representation from Equation (2.20) is not unique with respect to $A_t(\lambda)$. Even if it was constant over a certain time period, it would just be possible to identify the corresponding covariances up to a certain lag, which means that the spectral density remains unidentifiable because it is the Fourier transform of all covariances. However, taking the expectation of the periodogram over a segment of time, it can be shown that a convolution of $|A_t(\lambda)|^2$ with the Fejer kernel is uniquely determined from the process. Hence, not the spectrum itself, but an average is uniquely determined (see Priestley 1981, Section 11.2), which is obtained as expectation of a windowed periodogram over a certain time segment. This time segment has to be smaller than what Priestley calls the *bandwidth of stationarity.* If this bandwidth would be allowed to tend to infinity, the average spectrum from above would converge with an increasing time segment to the local spectrum, which would lead to the uniqueness of $|A_t(\lambda)|^2$. Yet, it is just characteristical for this class of processes, that the *bandwidth of stationarity* is not allowed to tend to infinity. This is the reason, why within Priestley's approach asymptotic considerations are not possible. (For further details about the evolutionary spectral density see Priestley 1981, pp. 821–855.)

2.2.2 Definition of local stationarity

In this context Dahlhaus (1997) has provided a decisive approach by introducing the concept of *locally stationary processes.* In what follows we will review the main ideas of this concept.

Dahlhaus' primary aim was to develop a model that, on the one hand, describes the phenomenon of local stationarity and, on the other hand, allows for asymptotic considerations concerning the estimation of the model parameters. Let us, for example, suppose that we observe a time-dependent AR(1) process of the form

$$X_t - \alpha(t)\, X_{t-1} = Z_t, \quad \{Z_t\} \text{ i. i. d. } \mathcal{N}(0, \sigma^2),$$

with $t = 1, \ldots, T$. The estimation of the coefficient function $\alpha(t)$ is performed by a parametric approach, for example by a second-degree polynomial. There are a multitude of classical methods to estimate the parameters ψ_0, ψ_1, and ψ_2 of such a function $\alpha_{\boldsymbol{\theta}}(t) := \psi_0 + \psi_1 t + \psi_2 t^2$. However, classical statements concerning the asymptotical behaviour of $\alpha_{\boldsymbol{\theta}}(t)$ (letting T tend to infinity) do not make sense, since for $T \to \infty$, it is $\alpha_{\boldsymbol{\theta}} \to \infty$, even though $\alpha_{\boldsymbol{\theta}}$ can be bounded for the observed time span.

To solve this problem, Dahlhaus uses an approach from the field of non-parametric regression. The time is rescaled on the interval $(0, 1]$. After this rescalation, an AR(1) process has the form

$$X_{t,T} - \alpha\Big(\frac{t}{T}\Big)\, X_{t-1,T} = Z_t.$$

Hence, growing T now coincides with growing information about the local structure of α. If α is constant over time, $X_{t,T}$ is independent of T, which leads to the classical asymptotic results for stationary time series. In order to expand this example to a more general class of locally stationary processes, we can use the rescalation in the spectral representation, i. e. we regard processes of the form

$$X_{t,T} = \int_{-\pi}^{\pi} A_{t,T}(\lambda) \exp(i\lambda t) \, \mathrm{d}Z(\lambda), \quad t = 1, \ldots, T, \, T > 0. \tag{2.21}$$

Yet, the time-dependent AR(1) process from our example does not have an exact, but only an approximate spectral representation of the form (2.21). That is why the definition of *local stationarity* has to be formulated a little bit more generally.

Definition 12
A sequence of stochastic processes $\{X_{t,T}\}_{1 \le t \le T}$ *is called locally stationary with transfer function* A°*, if there is a representation*

$$X_{t,T} = \int_{-\pi}^{\pi} A^\circ_{t,T}(\lambda) \exp(i\lambda t) \, \mathrm{d}Z(\lambda), \quad t = 1, \ldots, T, \, T > 0 \tag{2.22}$$

where the following holds:
a) $Z(\lambda)$ *is a complex-valued Gaussian process on* $[-\pi, \pi]$ *with* $\overline{Z(\lambda)} = Z(-\lambda)$, $\mathrm{E}\{Z(\lambda)\} = 0$, *and orthonormal increments, i. e.*

$$\mathrm{E}\{\mathrm{d}Z(\lambda_1) \, \overline{\mathrm{d}Z(\lambda_2)}\} = \eta(\lambda_1 + \lambda_2) \, \mathrm{d}\lambda_1 \, \mathrm{d}\lambda_2 \, ,$$

with the 2π*-extension of the Dirac-delta-function*

$$\eta(\lambda) = \sum_{j=-\infty}^{\infty} \delta(\lambda + 2\pi j) \, .$$

b) There is a positive constant K *and a continuous function* $A(u, \lambda)$ *on* $(0, 1] \times [-\pi, \pi)$*, which is* 2π*-periodic in* λ *with* $A(u, -\lambda) = \overline{A(u, \lambda)}$*, so that for all* $T > 0$

$$\sup_{t,\lambda} |A^\circ_{t,T}(\lambda) - A(t/T, \lambda)| \le K/T \tag{2.23}$$

holds.

The notation $\{X_{t,T}\}_{1\leq t\leq T}$ is used to underline that $\{X_{t,T}\}$ is triangular. Equation (2.23) is necessary to include the class of TVAR(p) processes in the class of locally stationary processes (see Dahlhaus 1996, pp. 145 et seqq.). These processes are described in Chapter 2.2.5.

Definition 12 uses an extension of the spectral representation of a stationary process (cf. Theorem 5 in Chapter 2.1). The difference between the spectral representation of the stationary process and that of the locally stationary process lies in the transfer function $A(u,\lambda)$, which depends on time and frequency and is defined on $(0,1]\times[-\pi,\pi]$. To ensure locally stationary behaviour of the process, it is necessary to state a few assumptions on the smoothness of A in u. These assumptions are expressed by the *total variation norm.*

In Definition 12 local stationarity is described by the assumption of continuity of $A(u,\lambda)$ in u. Thereby, Equation (2.23) ensures that the transfer function is also changing smoothly in time.

In what follows we give a simple example of a locally stationary process:

Example 1
Let $\{Y_t\}$ be a stationary process with spectral representation

$$Y_t = \int_{-\pi}^{\pi} \exp(i\lambda t)\, A(\lambda)\, \mathrm{d}Z(\lambda)$$

with $\mu = 0$, and $\sigma : (0,1] \to \mathbb{R}$ is a continuous function. Then

$$X_{t,T} := \sigma\Big(\frac{t}{T}\Big)\, Y_t$$

is a locally stationary process with $A^{\circ}_{t,T}(\lambda) = A(\frac{t}{T},\lambda) = \sigma(\frac{t}{T})\, A(\lambda)$.

Now we wish to derive an asymptotically well-defined spectrum. Therefore, we need the following definition:

Definition 13
For a univariate function f on an interval $[a,b]$ the total variation norm *is defined as*

$$TV_{[a,b]}(f) = \sup \left\{ \sum_{i=1}^{I} |f(a_i) - f(a_{i-1})| : a < a_0 < \ldots < a_I < b,\, I \in \mathbb{N} \right\}.$$

Following Neumann and von Sachs we make the following assumption (see Neumann and von Sachs 1997, p. 51):

Assumption 1
Function A in Definition 12 is such that
a) $\sup_u [TV_{[-\pi,\pi]}\{A(u,\cdot)\}] \le C_1 < \infty$,
b) $\sup_\lambda [TV_{[0,1]}\{A(\cdot,\lambda)\}] \le C_2 < \infty$,
c) $\sup_{u,\lambda} |A(u,\lambda)| \le \kappa_s < \infty$,
d) $\inf_{u,\lambda} |A(u,\lambda)| \ge \kappa_l$ *for some* $\kappa_l > 0$,
e) $\sup_u \{\sum_{s\in\mathbb{Z}} |\tilde{A}(u,s)|\} < \infty$, *where*

$$\tilde{A}(u,s) := (2\pi)^{-1} \int_{-\pi}^{\pi} A(u,\lambda)\, \exp(i\lambda s)\, \mathrm{d}\lambda$$

for $s \in \mathbb{Z}$ *and* $u \in (0,1]$.

Neumann and von Sachs have shown that under Assumption 1 we obtain the following Theorem:

Theorem 9
It is

$$\int_0^1 \int_{-\pi}^{\pi} |f_T(u,\lambda) - f(u,\lambda)|^2\, \mathrm{d}\lambda\, \mathrm{d}u = o_T(1), \tag{2.24}$$

where

$$f(u,\lambda) := |A(u,\lambda)|^2$$

is called evolutionary spectral density *and* $f_T(u,\lambda)$ *is the* Wigner-Ville spectrum.

For the proof see Neumann and von Sachs (1997, pp. 63–65). (For further details about the evolutionary spectral density see Priestley 1981, pp. 821–855.) The Wigner-Ville spectrum is a unique, time-dependent generalized spectrum that is based on the covariance function. For a locally stationary process $\{X_t\}_{1\le t\le T}$ and under the convention that

$A^{\circ}_{t,T}(\lambda) = A(0,\lambda)$ for $t < 1$ and $A^{\circ}_{t,T}(\lambda) = A(1,\lambda)$ for $t > T$ it has the form

$$f_T(u,\lambda) = \frac{1}{2\pi} \sum_{s=-\infty}^{\infty} \text{Cov}\left(X_{\lceil uT-s/2\rceil,T}, X_{\lceil uT+s/2\rceil,T}\right) \exp(-i\lambda s).$$

Here $\lceil u \rceil$ denotes the floor function of u.

Theorem 9 plays an important role concerning the uniqueness of the spectral representation from Equation (2.22). If there exists a connection as described in Equation (2.24) with a smooth function $A(u,\lambda)$, it follows from Theorem 9 that $|A(u,\lambda)|^2$ is asymptotically uniquely determined.

2.2.3 Local covariance estimation

The theoretical autocovariance of a locally stationary process has the form

$$\gamma(u,k) = \int_{-\pi}^{\pi} f(u,\lambda) \exp(i\lambda k)\, \mathrm{d}\lambda$$

which is equivalent to

$$\gamma(u,k) = \int_{-\pi}^{\pi} |A(u,\lambda)|^2 \exp(i\lambda k)\, \mathrm{d}\lambda.$$

In what follows we assume (again) $\mu(u) = 0$.

Dahlhaus has derived a decisive result concerning the estimation of the local partial autocorrelation (see Dahlhaus 2009):

Theorem 10
Let

$$\hat{\gamma}_T(u,k) := \frac{1}{b_T T} \sum_t K\left(\frac{u - (t+k/2)/T}{b_T}\right) X_{t,T} X_{t+k,T},$$

where K is a kernel with $K(x) = K(-x)$, $\int K(x)\,\mathrm{d}x = 1$, and $K(x) = 0$ for $x \notin [-1/2, 1/2]$ and bandwidth b_T.

Then

$$\sqrt{b_T T}(\hat{\gamma}_T(u_k, k) - \gamma(u_k, k))_{k=1,\ldots,d} \xrightarrow{\mathcal{D}} \mathcal{N}(0, \boldsymbol{\Sigma}) \quad \text{for} \quad T \to \infty$$

with

$$\boldsymbol{\Sigma}_{k,l} = \delta_{u_k,u_l} \int K^2(x)\,\mathrm{d}x \left[2\pi \int_{-\pi}^{\pi} \cos(\lambda k)\, \cos(\lambda l)\, f(u_k, \lambda)^2\,\mathrm{d}\lambda\, \kappa_4\, \gamma(u_k, k)\, \gamma(u_k, l)\right].$$

Using Parseval's formula (see for example Katznelson 2004, p. 37) and $\cos(\lambda k) = \frac{1}{2}[\exp(i\lambda k) + \exp(-i\lambda k)]$ the term in brackets is $\sum_{j=-\infty}^{\infty}[c(u_k, j-k-l) + c(u_k, j)\,c(u_k, j-k+l)] + \kappa_4\, c(u_k, k)\, c(u_k, l)$. This is the same as the usual asymptotic variance of the covariance-estimator (see Fuller 1996, p. 342, Theorem 6.4.1).

2.2.4 Local partial autocorrelation

As explained in Section 2.1.4, the partial autocorrelation $\phi(k)$ for a stationary process can be computed using the autocorrelation matrix. This motivates to define the following time-dependent version of the partial autocorrelation.

Definition 14
Let

$$\boldsymbol{P}_{t,T} := \begin{pmatrix} 1 & \gamma(t/T,1) & \gamma(t/T,2) & \ldots & \gamma(t/T,T-1) \\ \gamma(t/T,1) & 1 & \gamma(t/T,1) & \ldots & \gamma(t/T,T-2) \\ \gamma(t/T,2) & \gamma(t/T,1) & 1 & \ldots & \gamma(t/T,T-3) \\ \vdots & \vdots & & \ddots & \vdots \\ \gamma(t/T,T-1) & \gamma(t/T,T-2) & \gamma(t/T,T-3) & \ldots & 1 \end{pmatrix}$$

be the autocorrelation matrix at time t for a locally stationary process of length T. Then the local partial autocorrelation $\phi(t/T, k)$ *is defined as*

$$\phi(t/T, k) := \frac{\det \boldsymbol{P}^*_{t,k}}{\det \boldsymbol{P}_{t,k}},$$

where $\boldsymbol{P}_{t,k}$ is the $k \times k$ autocorrelation matrix at time t, and $\boldsymbol{P}^*_{t,k}$ is $\boldsymbol{P}_{t,k}$ with the last column replaced by

$$\begin{pmatrix} \gamma(t/T, 1) \\ \vdots \\ \gamma(t/T, k) \end{pmatrix}.$$

The asymptotical normality of $\hat{\phi}(k)$ (for the case of stationarity) can be derived from the multivariate normality of the empirical autocovariances. The behaviour of the covariance estimator described by Theorem 10 motivates for the computation of time-dependent partial autocorrelation estimators using the above defined covariance estimator. A formal proof showing the normality of the resulting estimator, which we suggest to name the *local partial autocorrelation* estimator should be part of future work. Here a possible ansatz could be to follow for example Serfling (2002, Theorem A, p. 122 et seqq.). This Theorem gives results concerning the application of multivariate functions to a vector of random variables. Under certain conditions, the asymptotic normality of the values from these functions can be deduced from the asymptotical normality of the random variables. In Section 3.2 we will suggest an approach to compute a preestimator of the time-dependent partial autocorrelations. This estimator will be used during the applications (Chapter 5) for choosing the order of the *time-varying autoregressive processes* that are fitted to the data. An introduction to the class of time-varying autoregressive processes is given in the next section.

2.2.5 TVAR

An important example of locally stationary processes is given by time-varying autoregressive processes (TVAR). (The proof of the local stationary of TVAR processes is provided in Dahlhaus 1996, p. 144). The TVAR models that we study are parameterized by a D-dimensional function $\boldsymbol{\theta}(u)$. They have the form

$$X_{t,T} + \sum_{j=1}^{p} \alpha_j\left(\frac{t}{T}\right) X_{t-j,T} = e_{t,T}, \quad t = 1, \ldots, T,\ T > 0. \qquad (2.27)$$

Here $e_{t,T}$ are independent normally distributed random variables with $\mu = 0$ and a constant standard deviation σ. The parameters are rescaled to the interval $(0, 1]$ and $\boldsymbol{\theta}(u) = (\sigma, \alpha_1(u), \ldots, \alpha_p(u))$ for $u \in (0, 1]$.[2]

The following Theorem is of great importance as without it $f(u, \lambda)$ cannot be interpreted as the spectral density of a TVAR process.

Theorem 11
Suppose that

$$\sum_{j=0}^{p} \alpha_j(u)\, z^j \neq 0$$

for all $|z| \leq 1 + c$ *with* $c > 0$ *uniformly in* u, $\alpha_0 = -1$, *and the coefficient functions* $\alpha_j(u)$ *are continuous on* $\mathbb{R}$. *Then* (2.27) *has a solution of the form* (2.22) *with*

$$A(u, \lambda) := \frac{\sigma}{\sqrt{2\pi}} \left(1 + \sum_{j=1}^{p} \alpha_j(u)\, \exp(-ij\lambda) \right)^{-1}$$

and time varying spectral density

$$f(u, \lambda) = \frac{\sigma^2}{2\pi} \left| \sum_{j=0}^{p} \alpha_j(u)\, \exp(i\lambda j) \right|^{-2}$$

(for the proof see Dahlhaus 1996, pp. 145 et seq).

[2] So it follows that $D = p + 1$.

3

Estimation

3.1 Maximum likelihood estimation with the Kullback-Leibler information divergence

In literature, there exists a set of different possibilities to estimate the parameters that specify a stationary ARMA model. In addition to the maximum likelihood and the least squares approach, which are used to fit ARMA models in the time domain, there also exist approaches to fit (stationary) ARMA models in the frequency domain. (For further details of the general approach see e. g. Schlittgen and Streitberg 2001, p. 376 et seq.) In what follows, we will recall a commonly used method for fitting stationary ARMA models in the frequency domain. The procedure will then be expanded to the case of locally stationary processes.

Let $\{X_t\}$ be an ARMA process as defined in Definition 10. The theoretical spectrum of such a process is given by Theorem 6. The model parameters can be estimated by fitting a function of type (2.16) to the observed periodogram $I(\lambda)$.[1] A systematical ansatz is given by the maximum likelihood approach. Usually, the exact likelihood function cannot be calculated easily. That is why we calculate the *Whittle likelihood.* Thereby we make use of the asymptotic distributions of the periodogram values for the Fourier frequencies. (See Brillinger 1981, p. 509 for further details.)

[1] The periodogram is the Fourier transform of the empirical autocovariances. For further details see Brockwell and Davis 1991, §10.1.

For the joint density of $I(\lambda_1), \ldots, I(\lambda_k), \ldots, I(\lambda_M)$ the Whittle likelihood is

$$g(I(\lambda_1), \ldots, I(\lambda_M)) = \prod_{k=1}^{M} \frac{1}{f(\lambda_k)} \exp\left(-\frac{I(\lambda_k)}{f(\lambda_k)}\right),$$

i.e. for a Fourier frequency $\lambda_k = \frac{k}{T}$ the periodogram is distributed asymptotically as an exponential variate with mean $f(\lambda_k)$. Moreover, $I(\lambda_1), \ldots, I(\lambda_k), \ldots, I(\lambda_M)$ are asymptotically independent (cf. Brillinger 1981, p. 509). Regarding the logarithmized density as a function of the unknown parameters, we get the loglikelihood function

$$l(\boldsymbol{\theta}) = -\sum_{k=1}^{M} \log f(\lambda_k) - \sum_{k=1}^{M} \frac{I(\lambda_k)}{f(\lambda_k)}, \tag{3.1}$$

for which we have to substitute $f(\lambda)$ by (2.16).

An estimator which maximizes (3.1) is called *Whittle estimator*. The maximization has to be done by an iterative approximation procedure. For AR processes the Whittle estimator corresponds to the *Yule-Walker estimator* (cf. Schlittgen and Streitberg 2001, p. 377). Another justification for the application of the Whittle likelihood is its correspondence to the (asymptotic) Kullback-Leibler divergence of two Gaussian processes with different spectral densities.

To understand the meaning of this correspondence we want to have a closer look at the Kullback-Leibler information divergence (see Kullback and Leibler 1951 or Kullback 1959 for details), which is an extension of the *entropy measure* (Shannon 1948).

Suppose that we have a T-dimensional random vector $\boldsymbol{X}$ and that the spectral density of $\boldsymbol{X}$ belongs to the family $\{f(\cdot;\boldsymbol{\psi}), \boldsymbol{\psi} \in \boldsymbol{\Psi}\}$. Then the Kullback-Leibler information divergence between $f(\cdot;\boldsymbol{\psi})$ and $f(\cdot;\boldsymbol{\theta})$ is defined as

$$d(\boldsymbol{\psi}|\boldsymbol{\theta}) = \Delta(\boldsymbol{\psi}|\boldsymbol{\theta}) - \Delta(\boldsymbol{\theta}|\boldsymbol{\theta}),$$

where

$$\begin{aligned}\Delta(\boldsymbol{\psi}|\boldsymbol{\theta}) &= \mathrm{E}_{\boldsymbol{\theta}}(-2\log f(\boldsymbol{X};\boldsymbol{\psi})) \\ &= \int_{\mathbb{R}^T} -2\log(f(\boldsymbol{x};\boldsymbol{\psi}))\, f(\boldsymbol{x};\boldsymbol{\theta})\,\mathrm{d}\boldsymbol{x}\end{aligned}$$

is the Kullback-Leibler index of $f(\cdot;\boldsymbol{\psi})$ relative to $f(\cdot;\boldsymbol{\theta})$. (In general it is $\Delta(\boldsymbol{\psi}|\boldsymbol{\theta}) \neq \Delta(\boldsymbol{\theta}|\boldsymbol{\psi})$.) With Jensen's inequality (see e. g. Rudin 1987, p. 62) we get

$$\begin{aligned}d(\boldsymbol{\psi}|\boldsymbol{\theta}) &= \int_{\mathbb{R}^T} -2\log\left(\frac{f(\boldsymbol{x};\boldsymbol{\psi})}{f(\boldsymbol{x};\boldsymbol{\theta})}\right) f(\boldsymbol{x};\boldsymbol{\theta})\,\mathrm{d}\boldsymbol{x} \\ &\geq -2\log\left(\int_{\mathbb{R}^T} \frac{f(\boldsymbol{x};\boldsymbol{\psi})}{f(\boldsymbol{x};\boldsymbol{\theta})} f(\boldsymbol{x};\boldsymbol{\theta})\,\mathrm{d}\boldsymbol{x}\right) \\ &= -2\log\left(\int_{\mathbb{R}^T} f(\boldsymbol{x};\boldsymbol{\psi})\,\mathrm{d}\boldsymbol{x}\right) \\ &= 0\end{aligned}$$

with equality if and only if $f(\boldsymbol{x};\boldsymbol{\psi}) = f(\boldsymbol{x};\boldsymbol{\theta})$.

Now suppose we observe data $X_1,\ldots,X_T$ with unknown parameters $\boldsymbol{\theta}$. The true model could be identified, if it was possible to compute the Kullback-Leibler information divergence between the respective potential models and the true model in each case. As this is not possible, we have to estimate the Kullback-Leibler information divergence and choose that model, for which the estimated Kullback-Leibler information divergence reaches its minimum value. Therefore, we assume that the true model and all the candidate models are Gaussian. (For further details see Brockwell and Davis 1991, pp. 302 et seqq.)

Let us now suppose that we have data $X_{1,T},\ldots,X_{T,T}$ to which we want to fit a locally stationary model with parameters $\boldsymbol{\theta} \in \Theta \subset \mathbb{R}^D$. Analogue to the stationary case above, we want to estimate the parameters by maximizing the likelihood. The processes are doubly-indexed to underline that we assume the observed sequence $x_t,\ldots,x_T$ is a realization from the triangular array

$$\begin{array}{lllll} X_{1,1} & & & & \\ X_{1,2} & X_{2,2} & & & \\ X_{1,3} & X_{2,3} & X_{3,3} & & \\ \vdots & \vdots & \vdots & \ddots & \\ X_{1,T} & X_{2,T} & X_{3,T} & \cdots & X_{T,T}. \end{array}$$

Let

$$\boldsymbol{\Sigma}_T(A,B) = \left\{ \int_{-\pi}^{\pi} \exp[i\lambda(r-s)]\, A^\circ_{r,T}(\lambda)\, \overline{B^\circ_{s,T}(\lambda)}\, \mathrm{d}\lambda \right\}_{r,s=1,\ldots,T}.$$

If the true process is locally stationary with transfer function A°, $\boldsymbol{\Sigma} = \boldsymbol{\Sigma}_T(A,A)$ is its true covariance matrix. For a Gaussian model with transfer function $A^\circ_{\boldsymbol{\theta}}$ we get the covariance matrix $\boldsymbol{\Sigma}_{\boldsymbol{\theta}} = \boldsymbol{\Sigma}_T(A_{\boldsymbol{\theta}}, A_{\boldsymbol{\theta}})$ and the maximum likelihood estimate

$$\hat{\boldsymbol{\theta}}_T := \arg\min_{\boldsymbol{\theta}\in\Theta} \mathcal{L}_T(\boldsymbol{\theta}),$$

where

$$\begin{aligned} \mathcal{L}_T(\boldsymbol{\theta}) &:= -\frac{1}{T}\text{Gaussian loglikelihood} \\ &= \frac{1}{2}\log 2\pi + \frac{1}{2T}\log\det \boldsymbol{\Sigma}_{\boldsymbol{\theta}} + \frac{1}{2T}\boldsymbol{X}'\boldsymbol{\Sigma}_{\boldsymbol{\theta}}^{-1}\boldsymbol{X} \end{aligned}$$

with $\boldsymbol{X} = (X_{1,T}\ldots X_{T,T})'$. Under certain regularity conditions $\hat{\boldsymbol{\theta}}_T$ will converge to

$$\boldsymbol{\theta}_\circ := \arg\min_{\boldsymbol{\theta}\in\Theta} \mathcal{L}(\boldsymbol{\theta}), \tag{3.10}$$

where

$$\mathcal{L}(\boldsymbol{\theta}) := \lim_{T\to\infty} E(\mathcal{L}_T(\boldsymbol{\theta}))$$

and $\lim_{T\to\infty} E$ is convergence in mean. If the model is correct, i.e. $A^\circ = A^\circ_{\boldsymbol{\theta}^*}$, it is $\boldsymbol{\theta}_\circ = \boldsymbol{\theta}^*$ (see Dahlhaus 2000, Theorem 2.3).

Suppose now that the true process fulfils the following assumption:

Assumption 2
$X_{1,T}, \ldots, X_{T,T}$ are realizations of a locally stationary process with transfer function A° where the corresponding A is bounded from below and has uniformly bounded derivative

$$\frac{\partial}{\partial u}\frac{\partial}{\partial \lambda}A.$$

$f(u,\lambda) = |A(u,\lambda)|^2$ denotes the time-varying spectral density of $\{X_{t,T}\}$.

This leads us to the asymptotic Kullback-Leibler information divergence for locally-stationary processes:

Theorem 12
Let Assumption 2 hold and the model consist of a locally stationary process with transfer function $A^\circ_{\boldsymbol{\theta}}$. Then with $f_{\boldsymbol{\theta}}(u,\lambda) = |A_{\boldsymbol{\theta}}(u,\lambda)|^2$ it is

$$\begin{aligned}\mathcal{L}(\boldsymbol{\theta}) &:= \lim_{T\to\infty} \mathrm{E}(\mathcal{L}_T(\boldsymbol{\theta})) \\ &= \frac{1}{4\pi}\int_0^1 \int_{-\pi}^{\pi} \left[\log 4\pi^2 f_{\boldsymbol{\theta}}(u,\lambda) + \frac{f(u,\lambda)}{f_{\boldsymbol{\theta}}(u,\lambda)}\right] \mathrm{d}\lambda\, \mathrm{d}u\end{aligned} \tag{3.11}$$

and

$$\mathcal{L}_T(\boldsymbol{\theta}) \to \mathcal{L}(\boldsymbol{\theta})$$

in probability.

For the proof see Dahlhaus (1996, pp. 149 et seq).

For being able to analyse the case that the model is not correct, Dahlhaus (2000) provides the asymptotic Kullback-Leibler information divergence for two locally stationary Gaussian processes $X_{1,T}\ldots X_{T,T}$ and $\tilde{X}_{1,T}\ldots\tilde{X}_{T,T}$ with densities g and $\tilde{g}$, and spectral densities $f = |A|^2$ and $\tilde{f} = |\tilde{A}|^2$ respectively:

$$\begin{aligned}\mathscr{D}(\tilde{f}, f) &= \lim_{T\to\infty} \frac{1}{T}\mathrm{E}_g \log\frac{g}{\tilde{g}} \\ &= \frac{1}{4\pi}\int_0^1 \int_{-\pi}^{\pi} \left[\log\frac{\tilde{f}(u,\lambda)}{f(u,\lambda)} + \frac{f(u,\lambda)}{\tilde{f}(u,\lambda)} - 1\right] \mathrm{d}\lambda\, \mathrm{d}u.\end{aligned} \tag{3.12}$$

(See Parzen 1983, pp. 230 et seqq. for a discussion in the stationary case.) It is a measure for the difference between the spectral densities of the two locally stationary Gaussian processes. As $\boldsymbol{\theta}_\circ$ from Equation (3.10) also minimizes $\mathscr{D}(\tilde{f}, f)$ from Equation (3.12), $\boldsymbol{\theta}_\circ$ is such that $f_{\boldsymbol{\theta}_\circ}$ is the best approximation of the true f in the sense of the above distance. It is the value to which the maximum likelihood estimate converges if the true process is not consistent with the fitted model (for the proof see Dahlhaus 2000).

The *contrast function* that we will use later in the estimation procedure has the form

$$\mathcal{L}(f_{\boldsymbol{\theta}}, f) = \frac{1}{4\pi} \int_0^1 \int_{-\pi}^{\pi} \left[\log f_{\boldsymbol{\theta}}(u, \lambda) + \frac{f(u, \lambda)}{f_{\boldsymbol{\theta}}(u, \lambda)} \right] \mathrm{d}\lambda \, \mathrm{d}u.$$

Thus, it is, except for a constant, the asymptotic Kullback-Leibler information divergence of a locally stationary process from Equation (3.11). Then, an *empirical contrast function* can be computed by

$$\mathcal{L}_T(f_{\hat{\boldsymbol{\theta}}}, J_T) = \frac{1}{4\pi T} \sum_{t=1}^{T} \int_{-\pi}^{\pi} \left[\log f_{\hat{\boldsymbol{\theta}}}(t/T, \lambda) + \frac{J_T(t/T, \lambda)}{f_{\hat{\boldsymbol{\theta}}}(t/T, \lambda)} \right] \mathrm{d}\lambda,$$

where the evolutionary spectral density is estimated by the preperiodogram

$$J_T(u, \lambda) = \frac{1}{2\pi} \sum_k X_{\lceil uT+(k+1)/2 \rceil, T} \, X_{\lceil uT-(k-1)/2 \rceil, T} \exp(-ik\lambda) \,,$$

whose definition is motivated by the convergence results (Equation 2.24) from the Wigner-Ville spectrum (Equation 2.2.2) (see Neumann and von Sachs 1997, p. 51).

3.2 Sieve estimation

As described in the previous Section we are interested in estimating the multivariate curve

$$\boldsymbol{\theta}(\cdot) = (\sigma, \alpha_1(\cdot), \ldots, \alpha_p(\cdot)) = (\theta^{(1)}(\cdot), \ldots, \theta^{(D)}(\cdot)).$$

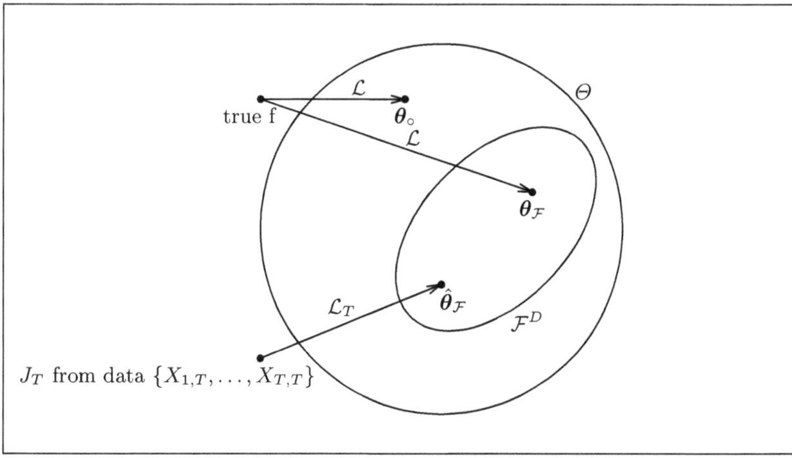

Figure 3.1: Estimation procedure on the finite-dimensional approximation space $\mathcal{F}^D$ (source: Van Bellegem 2003)

Instead of constructing an estimator by minimizing the empirical contrast function $\mathcal{L}_T(f_{\boldsymbol{\theta}}, J_T)$ over the class Θ of parameter curves (which would lead to serious numerical problems), we follow an approach that is based on the *method of sieves* suggested by Grenander (1981). The main idea of his approach is to perform the maximization of the likelihood within a subset of the parameter space. This subset, the *sieve*, is a sequence of approximating spaces. It is allowed to grow with the sample size. We approximate each of the time-varying parameters $\theta^{(i)}$ in a space of approximation $\mathcal{F}$, i. e. $\mathcal{L}_T$ is minimized over the product space $\mathcal{F}^D := \mathcal{F} \otimes \cdots \otimes \mathcal{F}$. This brings up the estimator $\hat{\boldsymbol{\theta}}_{\mathcal{F}}$.

Figure 3.1 illustrates the estimation procedure. $\boldsymbol{\theta}_\circ$ is given by minimizing the Kullback-Leibler information divergence $\mathcal{L}$ over the parameter space Θ of all spectral densities that follow the semiparametric structure with fixed D. $\mathcal{F}^D$ is a sieve on Θ, and the definition of $\boldsymbol{\theta}_{\mathcal{F}} \in \mathcal{F}^D$ is parallel to that of $\boldsymbol{\theta}_\circ$. Against it, if we observe data $\{X_{1,T}, \dots, X_{T,T}\}$, $\hat{\boldsymbol{\theta}}_{\mathcal{F}}$ results from the minimization of the empirical contrast function $\mathcal{L}_T$ over the space $\mathcal{F}^D$.

Van Bellegem and Dahlhaus (2006) have shown that if $\boldsymbol{\theta}_\circ$ and $\boldsymbol{\gamma}$ are both D-dimensional curves and we define

$$\|\boldsymbol{\theta}_\circ - \boldsymbol{\gamma}\|_2^2 := \sum_{i=1}^{D} \int_0^1 (\theta^{(i)}(u) - \gamma^{(i)}(u))^2 \, \mathrm{d}u,$$

we have

$$\mathrm{E}\|\boldsymbol{\theta}_\circ - \hat{\boldsymbol{\theta}}_{\mathcal{F}}\|_2 \lesssim \|\boldsymbol{\theta}_\circ - \boldsymbol{\theta}_{\mathcal{F}}\|_2 + c_{\boldsymbol{\theta}} \sqrt{\frac{D \dim(\mathcal{F})}{T}} + \mathcal{O}(\frac{1}{\sqrt{T}}),$$

where

$$\boldsymbol{\theta}_{\mathcal{F}} = \arg \min_{\boldsymbol{\theta} \in \mathcal{F}^D} \mathcal{L}(\boldsymbol{\theta}),$$

$$\hat{\boldsymbol{\theta}}_{\mathcal{F}} = \arg \min_{\boldsymbol{\theta} \in \mathcal{F}^D} \mathcal{L}_T(\boldsymbol{\theta}),$$

and $c_{\boldsymbol{\theta}}$ is a constant depending on $\boldsymbol{\theta}$. Here, $\lesssim$ means less or equal up to a finite constant independent of the parameters. While $c_{\boldsymbol{\theta}} \sqrt{\frac{D \dim(\mathcal{F})}{T}}$ is a measure of the bias, $\mathcal{O}(\frac{1}{\sqrt{T}})$ gives the variance of the risk.

Now we want to be more specific concerning the parameter space $\mathcal{F}^D$. We expand each component $\theta^{(i)}$ in an adequate basis $\{\varphi_j\}$, i. e.

$$\theta^{(i)} = \sum_{j=1}^{\infty} \psi_{ij} \, \varphi_j(u).$$

For the space $\mathcal{F}$, we take the m_i-dimensional linear space generated by $\varphi_1, \ldots, \varphi_{m_i}$. We set

$$\mathcal{N}_{D,T} = \{\boldsymbol{m} = (m_1, \ldots, m_D), m_j \in \mathcal{M}_T\}.$$

As explained above, the sieve is allowed to grow with the sample size. That is why the set $\mathcal{M}_T$ of dimensions is depending on T. For an illustration of this dependence see Van Bellegem (2003, Examples 2.1 to 2.4).

For each $\boldsymbol{m} = (m_1, \ldots, m_D)$ we define

$$\mathcal{F}^D = \mathcal{F}_{\boldsymbol{m}} := \mathcal{F}_{m_1} \otimes \cdots \otimes \mathcal{F}_{m_D}.$$

A number of different bases can be used for $\{\varphi_j\}$, examples are

polynomials:

$$\varphi_j(u) = u^j,\ j \geq 0,$$

trigonometric polynomials:

$$\varphi_0(u) = 1,\ \varphi_{2j}(u) = \cos(2j\pi u),\ \varphi_{2j+1}(u) = \sin(2j\pi u),\ j \geq 1,$$

splines:

$$\varphi_j(u) = u^i,\ \ j = 0, \ldots, r \ \text{ and } \ \varphi_{r+j}(u) = \max\{0, u - r_j\}^r,\ j = 1, \ldots, D.$$

Geman and Hwang have shown for very general settings that the method of sieves leads to consistent nonparametric estimators (Geman and Hwang 1982).

It remains the problem of finding a suitable $\hat{\boldsymbol{m}}$ among the set of possible dimensions $\mathcal{M}_T$. Van Bellegem and Dahlhaus (2006) show that

$$\mathrm{E}\|\boldsymbol{\theta}_\circ - \hat{\boldsymbol{\theta}}_{\hat{\boldsymbol{m}}}\|_2^2 \lesssim \inf_{\boldsymbol{m} \in \mathcal{M}_T} \{\|\boldsymbol{\theta}_\circ - \boldsymbol{\theta}_{\boldsymbol{m}}\|_2^2 + L_m \frac{D\boldsymbol{m}}{T}\} + \mathcal{O}(\frac{1}{T})$$

holds for some weights $L_{\boldsymbol{m}}$. The weights $L_{\boldsymbol{m}}$ can be constant. However, a non-constant $L_{\boldsymbol{m}}$ is needed to prevent the dimension of the models from growing too slowly. (For the details see Van Bellegem and Dahlhaus 2006, p. 728.)

To ensure that the expected value of the contrast function does not exceed certain limits (Van Bellegem and Dahlhaus 2006, Theorem 2) the functions φ_j have to meet particular assumptions (Van Bellegem and Dahlhaus 2006, Assumption 2). For example, for

$$\varphi_j(u) = \sqrt{2}\cos(2\pi j u), \quad j = 0, \ldots, m_{i-1} \tag{3.13}$$

these assumptions hold. (For further examples of models that fulfil these assumptions see e. g. Birgé and Massart 1998, Barron, Birgé, and Massart 1999, and Comte 2001.)

The model that we want to fit to an empirical series is the TVAR(p) model described in Equation (1.3). The spectral density of a process from that class has the following form:

$$f_{\boldsymbol{\theta}}(u,\lambda) = \frac{\sigma^2}{2\pi} \frac{1}{|\sum_{j=0}^{p} \alpha_j(u)\, \exp(i\lambda j)|^2}$$

(see Equation 11). From this and Kolmogorov's formula it follows:

$$\begin{aligned}
&\mathcal{L}_T\left(f_{\boldsymbol{\theta}}, J_T\right) \\
&= \frac{1}{2T}\sum_{t=1}^{T}\left[\log\sigma^2 + \frac{1}{\sigma^2}\left\{\left(\boldsymbol{\Sigma}_{t,T}\,\boldsymbol{\alpha}\left(\frac{t}{T}\right) + \boldsymbol{C}_{t,T}\right)'\right.\right. \\
&\quad \left.\left.\boldsymbol{\Sigma}_{t,T}^{-1}\left(\boldsymbol{\Sigma}_{t,T}\,\boldsymbol{\alpha}\left(\frac{t}{T}\right) + \boldsymbol{C}_{t,T}\right) + c_T\left(\frac{t}{T},0\right) - \boldsymbol{C}_{t,T}'\boldsymbol{\Sigma}_{t,T}^{-1}\boldsymbol{C}_{t,T}\right\}\right]
\end{aligned}$$

(Van Bellegem and Dahlhaus 2006, pp. 730 et seq.) with

$$\begin{aligned}
\boldsymbol{\alpha}\left(\frac{t}{T}\right) &= \left(\alpha_1\left(\frac{t}{T}\right),\ldots,\alpha_p\left(\frac{t}{T}\right)\right)', \\
c_T\left(\frac{t}{T},j\right) &= \int_{-\pi}^{\pi} J_T\left(\frac{t}{T},\lambda\right)\exp(i\lambda j)\,\mathrm{d}\lambda = X_{\lceil t+(j+1)/2\rceil}\,X_{\lceil t-(j-1)/2\rceil}, \\
\boldsymbol{C}_{t,T} &= \left(c_T\left(\frac{t}{T},1\right),\ldots,c_T\left(\frac{t}{T},p\right)\right)', \\
\boldsymbol{\Sigma}_{t,T} &= \left\{c_T\left(\frac{t}{T},j-k\right)\right\}_{j,k=1,\ldots,p}.
\end{aligned}$$

The model selection procedure then consists of the following two steps:

a) On each space $\mathcal{F}_{\boldsymbol{m}}$ compute

$$\hat{\boldsymbol{\theta}}_{\boldsymbol{m}} = \arg\min_{\boldsymbol{\theta}\in\mathcal{F}_{\boldsymbol{m}}}\{\mathcal{L}_T(f_{\boldsymbol{\theta}}, J_T)\}$$

for $\boldsymbol{m}\in\mathcal{N}_{D,T}$.

b) Then choose $\hat{\boldsymbol{m}}$ among $\left\{\hat{\boldsymbol{\theta}}_{\boldsymbol{m}} : \boldsymbol{m}\in\mathcal{N}_{D,T}\right\}$ such that

$$\hat{\boldsymbol{m}} = \arg\min_{\boldsymbol{m}\in\mathcal{N}_{D,T}}\{\mathcal{L}_T(f_{\hat{\boldsymbol{\theta}}_{\boldsymbol{m}}}, J_T) + \mathrm{pen}(\boldsymbol{m})\}.$$

We then get the sieve estimator

$$\hat{\boldsymbol{\theta}} = \hat{\boldsymbol{\theta}}_{\hat{\boldsymbol{m}}}.$$

A penalty function is necessary to ensure the choice of a parsimonious model in the model selection procedure. We make use of the following penalty function:

$$\text{pen}(\boldsymbol{m}) = c_3 \frac{d_m}{T} + c_4 \frac{d_m(1 + L_{\boldsymbol{m}})}{T} ||\boldsymbol{\Sigma}||^2_{\text{spec}},$$

where c_3 and c_4 are positive coefficients. (For the choice of c_3 and c_4 see Van Bellegem and Dahlhaus 2006, p. 740.) $||\boldsymbol{\Sigma}||_{\text{spec}}$ is the spectral norm of the covariance matrix $\boldsymbol{\Sigma}$. It is defined as $||\boldsymbol{\Sigma}||_{\text{spec}} := \max\{\sqrt{\lambda} : \lambda \, \text{eigenvalue of } \boldsymbol{\Sigma}'\boldsymbol{\Sigma}\}$. As the covariance matrix is symmetric, this corresponds to the computation of the largest eigenvalue.

The implementation of the procedure requires the preestimation of $||\boldsymbol{\Sigma}||_{\text{spec}}$. This can be obtained by computing the largest eigenvalue of the matrix $\hat{\boldsymbol{\Sigma}}_{s,t}$ with elements

$$\hat{c}_T\left(\frac{s+t}{2T}, |s-t|\right) I(|s-t| \leq M).$$

$\hat{c}_T$ denotes the smoothing of $c_T(u, j)$ with respect to u, where $c_T(u, j)$ is a preestimator of

$$c(t/T, j) = \int \exp(i\lambda j)\, f(t/T, \lambda)\, \mathrm{d}\lambda\, .$$

In our simulations and applications the smoothing is done by a Nadaraya-Watson kernel estimator, which leads to

$$\hat{c}_T(t/T, j) = \frac{\sum_{s=1}^{T} c(t/T, j)\, K(\frac{t-s}{b})}{\sum_{s=1}^{T} K(\frac{t-s}{b})}$$

with a rectangular kernel function K and the bandwidth b.

For the choice of M we follow Van Bellegem and von Sachs (Van Bellegem and von Sachs 2008; Van Bellegem and von Sachs 2004). They suggest to compute the covariance matrix for a large value of M and to examine the behaviour of the off-diagonals. When an abrupt decreasing trend is observed in several off-diagonals, they recommend clipping the matrix right before this abrupt decrease. However, as our simulation results show, their recommendation to choose $[\log_2 T]$ for the computation of the bandwidth does not result in choosing an appropriate $\hat{\boldsymbol{m}}$. That is why we choose $b_T = T$ in our simulations and applications which provides better results. However, for the choice of M (and the order p as described next), we recommend to choose $b_T < T$.

For the choice of p, we also need a preestimator of the time-dependent partial autocorrelation function. For lag s at time t we define this preestimator as

$$\hat{\phi}(s,t) := \frac{\det \hat{\boldsymbol{C}}^*_{s,t}}{\det \hat{\boldsymbol{C}}_{s,t}}$$

with

$$\hat{\boldsymbol{C}}_{s,t} := \begin{pmatrix} \hat{c}_T(t/T,1) & \hat{c}_T(t/T,2) & \dots & \hat{c}_T(t/T,s) \\ \hat{c}_T(t/T,2) & \hat{c}_T(t/T,1) & \dots & \hat{c}_T(t/T,s-1) \\ \vdots & & \ddots & \vdots \\ \hat{c}_T(t/T,s) & \hat{c}_T(t/T,s-1) & \dots & \hat{c}_T(t/T,1) \end{pmatrix},$$

where $\hat{\boldsymbol{C}}^*_{s,t}$ is $\hat{\boldsymbol{C}}_{s,t}$ with the last columns replaced in the same way as for the autocorrelation matrix in Chapter 2. This definition is motivated by the connection of the autocovariances and the partial autocorrelations for stationary processes, which has been recalled in Section 2.1.4. We recommend to compute the $\hat{\phi}(s,t)$ for different t and until $s = M$ for M chosen as described above. We examine for which lags we observe

$$|\hat{\phi}(s,t)| > \frac{1}{\sqrt{b_T T}}.$$

This boundary is motivated by the corresponding boundaries for the stationary case. We investigate the behaviour of the local partial autocorrelation estimator for different times t and choose p such that

$$|\hat{\phi}(s,t)| < \frac{1}{\sqrt{b_T T}} \text{ for all } t \in \{1, \dots, T\} \text{ and all } s > p.$$

4

Forecasting

4.1 Prediction in the case of stationarity

As explained earlier, the main objective of this work is to compute predictions, more precisely, the deviation of a forecasting approach for locally stationary processes. But before we present our approach to forecast time series using locally stationary processes, we want to investigate the problem of predicting the future values of a stationary process, i. e. of predicting the values $\{X_t, t \geq T+1\}$ in terms of $\{X_1, \ldots, X_T\}$ for a stationary process $\{X_t\}_{t \in \mathbb{Z}}$.

Let $\{x_t\}_{t=1,\ldots,T}$ be a section from a realization of the stochastic process $\{X_t\}_{t \in \mathbb{Z}}$. In time series analysis, *predicting* generally means utilizing the observations taken at or before time T to forecast the subsequent behaviour of $\{X_t\}$. A nearby approach is to interpret the conditional expectation $\mathrm{E}(X|\mathcal{G})$ as prediction of X by means of the information $\mathcal{G}$ (cf. for example Meintrup and Schäffler 2005, pp. 223 et seq.). The motivation is stated in the following Theorem:

Theorem 13
Let $X \in L^2(\Omega, \mathcal{A}, P)$ and $\mathcal{G} \subset \mathcal{A}$ be a sub-σ-algebra. Then the function

$$h(Y) = \mathrm{E}((X-Y)^2),\ Y \in L^2(\Omega, \mathcal{G}, P|\mathcal{G})$$

takes a minimum in $X_0 := \mathrm{E}(X|\mathcal{G})$.

For the proof see again Meintrup and Schäffler (2005), pp. 223 et seq. Here, $P|\mathcal{G}$ means that the probability measure P is constrained to the sub-σ-algebra $\mathcal{G}$ (it is not the conditional probability).

However, the determination of projections on $\mathcal{G}$ is usually very difficult. (See for example Brockwell and Davis 1991, Section 2.7 for details.)
Yet, if $Y_1, \ldots, Y_T \in L^2$, it is relatively easy to compute the projection of X on the closed span[1] $\overline{sp}\{1, Y_1, \ldots, Y_T\} \subseteq \mathcal{G}(Y_1, \ldots, Y_T)$, as we can write

$$P_{\overline{sp}\{1,Y_1,\ldots,Y_T\}}(X) = \sum_{i=0}^{T} \alpha_i Y_i, \quad Y_0 = 1, \tag{4.1}$$

where $\alpha_0, \ldots, \alpha_T$ satisfy

$$\left\langle \sum_{i=0}^{T} \alpha_i Y_i, Y_j \right\rangle = \langle X, Y_j \rangle, \quad j = 0, 1, \ldots, T,$$

or equivalently

$$\sum_{i=0}^{T} \alpha_i \mathrm{E}(Y_i Y_j) = \mathrm{E}(X Y_j), \quad j = 0, 1, \ldots, T.$$

The projection theorem[2] guarantees that a solution $(\alpha_0, \ldots, \alpha_T)$ exists. A solution inserted in Equation (4.1) gives the required projection, the *best linear predictor* of X in terms of $1, Y_1, \ldots, Y_T$. Since it is a projection of X onto a subspace of $\mathcal{G}(Y_1, \ldots, Y_T)$, it can never have a smaller mean square error than the conditional expectation. Nevertheless, it is of great importance for the following reasons:

- it is easier to calculate,
- it depends only on the first and second order moments,
- if $(X, Y_1, \ldots, Y_T)'$ has a multivariate normal distribution, the best linear predictor corresponds to the conditional expectation (cf. for example Doob 1953, p. 390).

[1] The closed span of any subset of a Hilbert space is the smallest closed subspace of this Hilbert space which contains each element of the subset.
[2] For recalling the projection theorem and the corresponding proof cf. for example Pourahmadi 2001, p. 312.

In what follows we will give a more general definition of best linear predictors:

Theorem 14
If $X \in L^2$ and $Y_\lambda \in L^2$ for all $\lambda \in \Lambda$, then the best linear predictor of X in terms of $\{Y_\lambda, \lambda \in \Lambda\}$ is that element of $\overline{sp}\{Y_\lambda, \lambda \in \Lambda\}$ that has the smallest distance from X. With the projection theorem this is $P_{\overline{sp}\{Y_\lambda, \lambda \in \Lambda\}}(X)$.

For ARMA models, this leads to the following theorem:

Theorem 15
The optimal h-step-prediction $\hat{X}_{t,h}$ of a stationary and invertible ARMA(p,q) process $\alpha(L)X_t = \beta(L)e_t$ with $\mathrm{E}(e_t) = 0$ is given by the following interpretation of the model equation

$$X_{t+h} = \alpha_1 X_{t+h-1} + \ldots + \alpha_p X_{t+h-p} + e_{t+h} - \beta_1 e_{t+h-1} - \ldots - \beta_q e_{t+h-q} :$$

1. *$X_t, X_{t-1}, \ldots$ correspond to the actual observations.*
2. *The unobserved variables $X_{t+h}, X_{t+h-1}, \ldots, X_{t+1}$ are substituted by the optimal predictions $\hat{X}_{t,h}, \hat{X}_{t,h-1}, \ldots, \hat{X}_{t,1}$.*
3. *$e_t, e_{t-1}, \ldots$ correspond to the prediction errors $X_t - \hat{X}_{t-1,1}, X_{t-1} - \hat{X}_{t-2,1}, \ldots$ of the optimal 1-step-predictions.*
4. *$e_{t+h}, e_{t+h-1}, \ldots, e_{t+1}$ are replaced by zero, which is their expectation.*

For the proof see for example Schlittgen and Streitberg (2001), p. 215.

4.2 Approaches to forecast time series using TVAR processes

Below we propose three procedures, which are natural generalizations of the above Theorem, for forecasting time series using TVAR(p) processes. In the style of Van Bellegem and von Sachs (2002)[3] we make use of the following notation, to forecast the h values of an observed process: The observed variables are denoted by $X_{1,T}, \ldots, X_{T-h,T}$. Then the functions

[3] The work is focused on developing a forecasting procedure for the simple model from Example 1.

of the coefficients can be estimated on the interval $(0\,;\,1-\frac{h}{T}]$, whereas the values outside this interval have to be obtained by extrapolation.

Approach 1
The first approach is an adaption of the linear ansatz for stationary processes from above, where the constant coefficients are replaced by time-varying coefficient functions. For each forecasting step the values of the coefficients are computed for the corresponding prospective time point.

We estimate the parameter vector $\boldsymbol{\theta}(\frac{t}{T}) = (\sigma, \alpha_1(\frac{t}{T}), \ldots, \alpha_p(\frac{t}{T}))$ using

$$\vartheta_i(u) = \sum_{j=1}^{d_{m_i}} \psi_{ij}\,\varphi_j(u), \quad i = 1, \ldots, p,$$

where

$$\varphi_j(u) = \sqrt{2}\cos(2\pi(j-1)\zeta u), \quad j = 1, \ldots, d_{m_i} \tag{4.2}$$

and $u = \frac{t}{T}$ with $u \in (0; \frac{T-h}{T}]$, i. e. we estimate ψ_{ij} for every $i = 1, \ldots, p$ and $j = 1, \ldots, d_{m_i}$ using the same approach as Van Bellegem and Dahlhaus (2006) but with the difference that u is multiplied by a constant ζ, $\zeta \in (0; 1]$. The reason for this is the following: Continuous functions can be approximated uniformly by linear combinations of functions of the form in Equation (3.13) (*Fourier approximation*, see for example Brockwell and Davis 1991, §2.8). But this is only possible for functions $\theta_i(u)$, $u \in [0; 1]$, for which $\theta_i(0) = \theta_i(1)$ holds. As we want to be able to approximate functions, for which this is not the case, we expand the class of possible approximation-functions by using ζ as scale factor. $d_{m_1}, \ldots, d_{m_p}$ are chosen during the semiparametric estimation procedure.

We define the *h-range-predictor* $\hat{X}_{T-h}(k)$, $k = 1, \ldots, h$ of a TVAR(p) process as

$$
\begin{aligned}
\hat{X}_{T-h}(1) &= -\hat{\alpha}_1\Big(\frac{T-h+1}{T}\Big)\, X_{T-h} - \cdots \\
&\quad -\hat{\alpha}_p\Big(\frac{T-h+1}{T}\Big)\, X_{T-h-(p-1)} \\
\hat{X}_{T-h}(2) &= -\hat{\alpha}_1\Big(\frac{T-h+2}{T}\Big)\, \hat{X}_{T-h}(1) - \cdots \\
&\quad -\hat{\alpha}_p\Big(\frac{T-h+2}{T}\Big)\, X_{T-h-(p-2)} \\
&\vdots \\
\hat{X}_{T-h}(h) &= -\hat{\alpha}_1(1)\, \hat{X}_{T-h}(h-1) - \cdots - \hat{\alpha}_p(1)\, \hat{X}_{T-h-1}(h-p)
\end{aligned}
$$

with forecasting horizon h. The forecasting procedure is shown in Figure 4.1.

Approach 2

In the second approach we make use of constant coefficients for the different forecasting steps. Such an approach could be preferable compared to the first approach for two reasons. It is easier to compute forecasts with constant coefficients and, if there is no reason for assuming that the coefficient functions will develop in the same manner as they did during the observed time, it is more reasonable.

The computation of forecasts in this case is analogous to Approach 1, with the difference that $\hat{\boldsymbol{\theta}}$ is not extrapolated, but left constant as $\hat{\boldsymbol{\theta}}(\frac{T-h}{T})$ during different forecasting steps.

Approach 3

For nonlinear stochastic processes, in contrast to the linear case, multi-step-forecasting in the form of Approach 1 is not optimal any more (cf. Tong 1993, Chapter 6). We want to evaluate, if we can achieve an advancement concerning our predictions if we restrict the proceeding to one-step-forecasting. Nevertheless, we want to compute predictions for several steps. In the third approach, for a certain prediction step, we therefore treat the predictions of preceding steps as if they were observations.

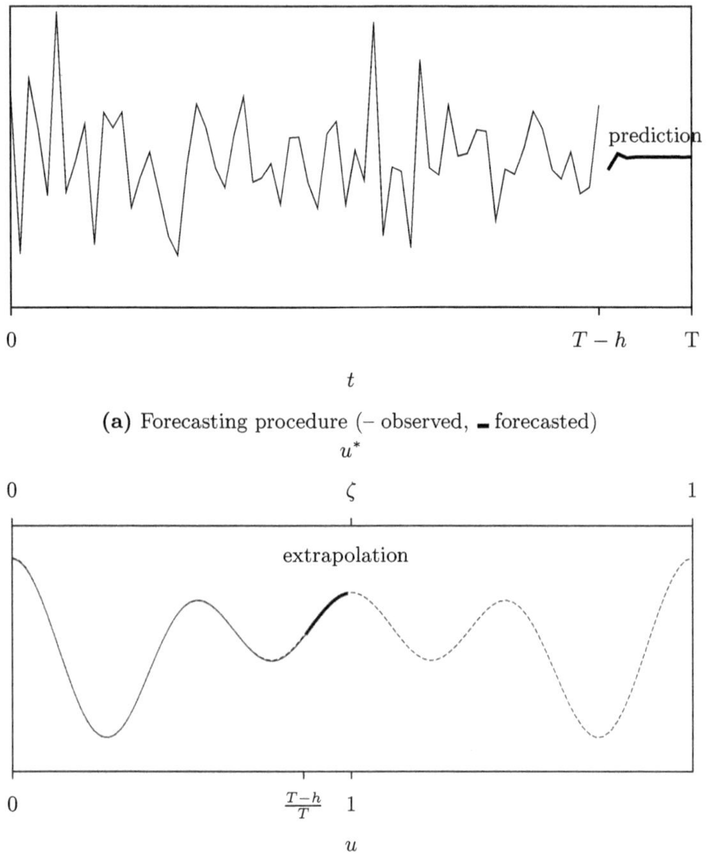

(a) Forecasting procedure (– observed, ▬ forecasted)

(b) Extrapolation of the coefficient functions (– approximated, ▬ extrapolated, -- true coefficient function)

Figure 4.1: Forecasting procedure and extrapolation of the coefficient functions ($u^* := \zeta u$)

We compute h one-step-ahead-forecasts using h different coefficients, i. e. we set $k = 1$ and define the k-step predictor $\hat{X}_{T-k}(k)$ of a TVAR(p)-process as

$$\begin{aligned}\hat{X}_{T-k}(k) = &-\hat{\theta}_1^{*,k}\left(\frac{T-h+k}{T}\right)\hat{X}_{T-k}(k-1)\cdots\\ &-\hat{\theta}_p^{*,k}\left(\frac{T-h+k}{T}\right)\hat{X}_{T-k-(p-1)}(k-p)\end{aligned}$$

where

$$\hat{X}_{\bullet}(s) = X_{\bullet} \quad \text{for} \quad s < 1$$

with

$$\hat{\theta}_i^{*,k}(u) = \sum_{j=1}^{d_{m_i}} \hat{\psi}_{ij}\,\varphi_j^{*,k}(u), \quad i = 1, \ldots, p,$$

where

$$\varphi_j^{*,k}(u) = \sqrt{2}\cos(\pi(j-1)\frac{u}{\zeta}), \quad j = 1, \ldots, d_{m_i}$$

with $u \in (0, \frac{T-h+k}{T}]$. We estimate the ψ_{ij} again using the x-values from the last estimation plus the last computed forecast. We set $k = k + 1$ and repeat this until $k = h$.

4.3 Iterative stages in the selection of a model

Figure 4.2 summarizes the iterative approach to model building for forecasting that is used in this work. It is an extension of the Box-Jenkins procedure (Box and Jenkins 1976). A more detailed description of the different steps is given in the following listing:

(1) From the interaction of theory and practice the class of TVAR(p) processes is chosen. If the time series has a trend, the trend is estimated and removed. Trend estimation is a well studied problem, but it is not in the focus of this work.

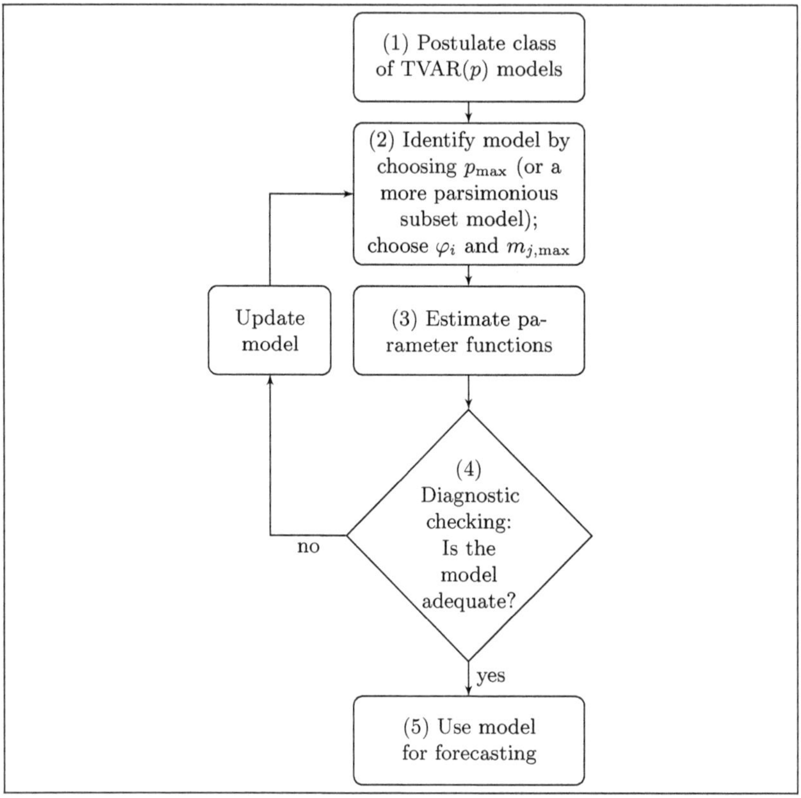

Figure 4.2: Stages in the iterative approach to model building

(2) For arbitrary p, this class is to extensive to be conveniently fitted to data, so one has to choose a maximum order $p_{\max}$ or a more parsimonious subset model, which can be done with the help of the local version of the partial autocorrelation function that is suggested in Section 3.2. Besides, the practitioner has to choose a suitable $m_{j,\max}$ and the functions φ_i for the approximation procedure. And he has to compute $\hat{\Sigma}$ until a large Lag (e. g. 10) for the choice of M. Then $c_T\big(\frac{t}{T}, j\big) = X_{[t+(j+1)/2]}X_{[t-(j-1)/2]}$ can be computed. After choosing a suitable bandwidth and a smoothing method, the smoothed $\hat{c}_T$ can be provided, which is needed in the estimation procedure.

(3) The model is fitted to data and its parameters are estimated. Therefore, the spectrum of the chosen TVAR process $f_{\boldsymbol{\theta}(u)}$ has to be computed. Then one can calculate the ψ_{ij} $(i = 1, \ldots, m_j, \quad j = 1, \ldots, p)$ such, that the distance between $f_{\boldsymbol{\theta}(u)}$ and the non-parametric estimator J_T of the true spectrum f is minimized. After that the m_j are chosen for the minimum value of a penalized version of the contrast function, i. e. the m_j are chosen such, that

$$\begin{aligned}\hat{\boldsymbol{m}} &= (\hat{m}_1, \hat{m}_2, \ldots, \hat{m}_j, \ldots, \hat{m}_p) \\ &= \arg\min_{\boldsymbol{m}} \{\mathcal{L}_T(f_{\hat{\boldsymbol{\theta}}_m}, J_T) + \operatorname{pen}(\boldsymbol{m})\}.\end{aligned}$$

These estimations are calculated for different ζ. Then, it is chosen that ζ, for which the mean square errors are minimized.

(4) Diagnostic checks are applied in order to uncover possible lack of fit and diagnosing the cause. If any inadequacy is found, the iterative cycle of identification, estimation, and diagnostic checking is repeated until a suitable representation is found.

4.4 Simulations

4.4.1 Course of action

In what follows, the performance of the prediction approaches is evaluated on three specific models, which are examples of stationary process. Each of the three models is a TVAR(2) process, i. e. the stochastic process from Equation (1.3) with $p = 2$. For this class of processes, we examine three different parameter constellations. The first constellation uses the trigonometric functions given in Equation (4.2) with $\zeta = 0.5$, the time-dependent coefficients having the form

$$\alpha_1(u) = \sqrt{2}\begin{pmatrix}0.2 & -0.1 & 0.1 & 0.2 & 0.3\end{pmatrix}\begin{pmatrix}1 \\ \cos(\pi u) \\ \cos(2\pi u) \\ \cos(3\pi u) \\ \cos(4\pi u)\end{pmatrix}$$

and

$$\alpha_2(u) = \sqrt{2} \cdot 0.3$$

respectively. Thereby, the vector $(0.2, -0.1, 0.1, 0.2, 0.3)$ and the constant 0.3 are arbitrarily chosen. In what follows this will be referred to as the *TRIG1* case.

The second constellation (*TRIG2*) makes use of the time-dependent coefficients

$$\alpha_1(u) = -1.8\cos(1.5 - \cos(4\pi u))$$

and

$$\alpha_2(u) = 0.81.$$

In the third constellation (*LEG*) Legendre polynomials are used. A plot of the first five Legendre polynomials can be found in Figure A.1. Therefore, the coefficients

$$\alpha_1(u) = \begin{pmatrix} 0.2 & -0.1 & 0.1 & 0.2 & 0.3 \end{pmatrix} \begin{pmatrix} 1 \\ u \\ 0.5(3u^2 - 1) \\ 0.5(5u^3 - 3u) \\ 0.125(35u^4 - 30u^2 + 3) \end{pmatrix}$$

and

$$\alpha_2(u) = 0.3$$

are chosen. In all three cases the standard deviation of the innovation process is set to 1. Typical realizations of the respective processes are given in Figure 4.3.

In the estimation procedure we use trigonometric sieves, i. e. we use functions of the form of Equation (4.2), which means that the coefficients of the first process have the exact same form as the sieves we use in the estimation procedure, whereas the coefficients of the second and third process cannot be written as a finite linear combination of trigonometric functions, in other words we are dealing with a misspecified case. Each of the Monte Carlo simulations is based on 1000 generations of the respective time-varying autoregressive process of sample sizes $T = 64$, $T = 256$, and $T = 1024$. For each of these realizations we compute forecasts following the forecasting procedures described above. In order to compare the estimation error for the different forecasting procedures, we compute the mean square error (MSE) for each of the 1000 generations.

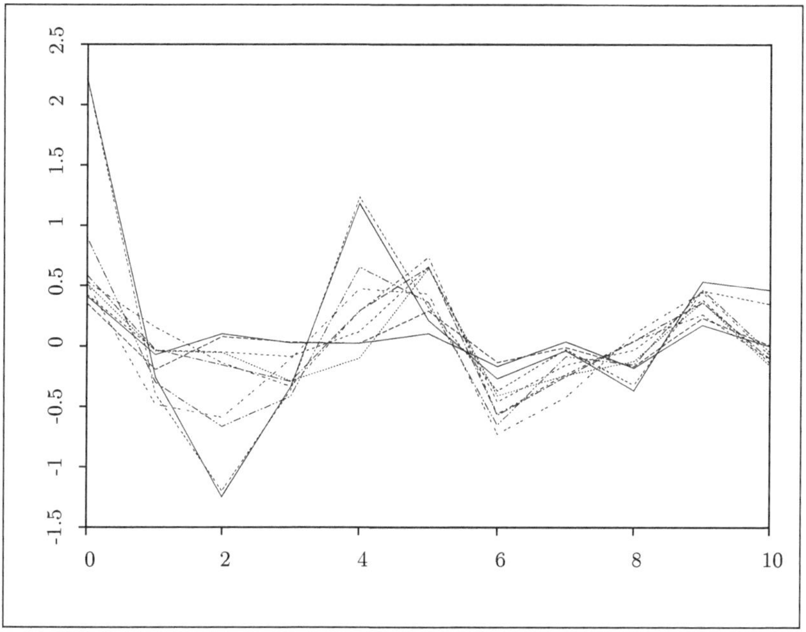

Figure 4.4: Ten off-diagonals of the estimated covariance matrix

The selection of M, for which we have preestimated the covariance matrix Σ, is provided as described in Section 3.2. Figure 4.4 shows 10 different off-diagonals for lags 0 to 10 of a TVAR process simulated on the basis of Legendre polynomials. It shows that there is a strong decrease between lag 4 and 5. For this reason we choose $M = 4$, i. e. we only use the main diagonal and the first four off-diagonals of the covariance matrix for the estimation of $||\Sigma||_{\text{spec}}$.

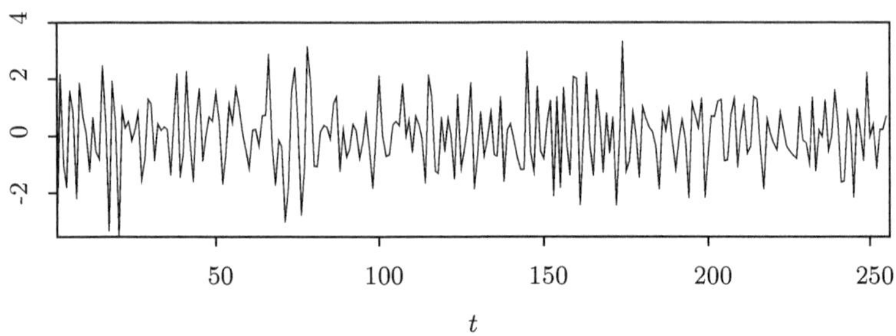

(a) TRIG1 case

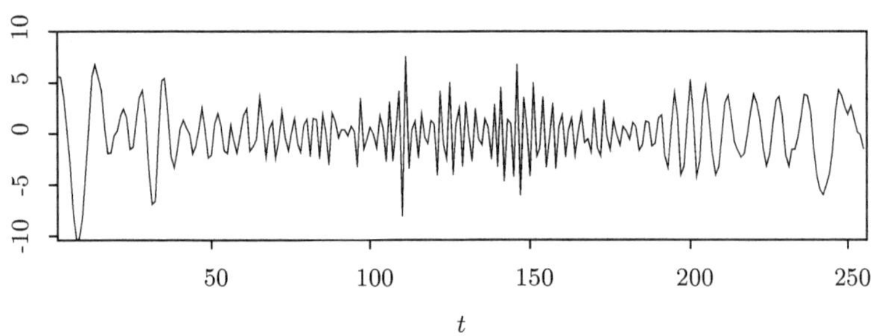

(b) TRIG2 case

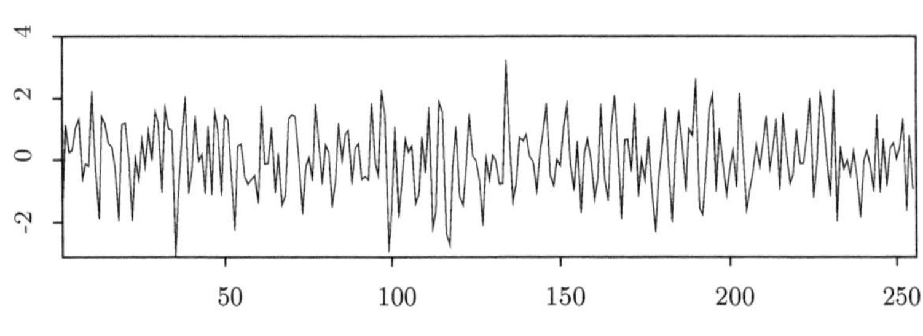

(c) LEG case

Figure 4.3: Realizations of TVAR(2) processes

4.4.2 Results

Table 4.1 gives the results of the Monte Carlo simulations described above. Thereby, solely TVAR(2) processes where fitted. The selection of the polynomial degrees are compared for different sample sizes and polynomials used in the selection procedure. Here $\bar{d}_{m_1}$ and $\bar{d}_{m_2}$ are the mean degrees selected. Numbers in parentheses are the empirical standard deviations. Values smaller one appear since the procedure sometimes selects the order $p = 1$. d_{m_2} is then set to zero.

sample size	$T = 64$		
(true) polynomials	TRIG1	TRIG2	LEG
$\bar{d}_{m_1}$	0.341	1.054	0.099
	(0.791)	(2.055)	(0.299)
$\bar{d}_{m_2}$	0.75	0.884	0.902
	(0.764)	(1.997)	(0.301)
sample size	$T = 256$		
(true) polynomials	TRIG1	TRIG2	LEG
$\bar{d}_{m_1}$	0.485	1.03	0.043
	(0.674)	(0.883)	(0.203)
$\bar{d}_{m_2}$	0.711	0.375	0.99
	(0.454)	(0.484)	(0.109)
sample size	$T = 1024$		
(true) polynomials	TRIG1	TRIG2	LEG
$\bar{d}_{m_1}$	4.988	2.97	1.576
	(0.219)	(0.318)	(1.228)
$\bar{d}_{m_2}$	1	0.996	1.001
	(0)	(0.190)	(0.0316)

Table 4.1: Results of the model selection procedure for simulations based on 1000 generations of a TVAR(2) process of sample sizes $T = 64$, $T = 256$, and $T = 1024$

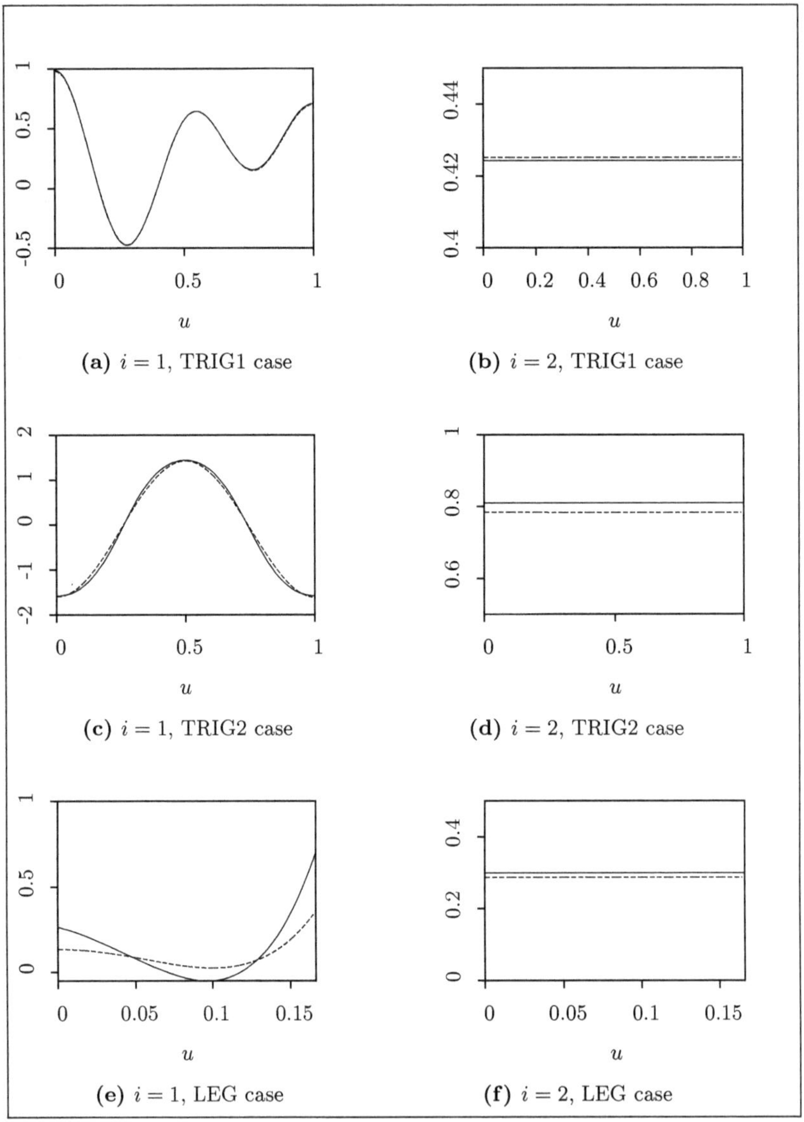

Figure 4.5: Time-dependent coefficient estimations based on the penalized likelihood method (— true curve α_i, - - - mean estimated curve $\bar{\hat{\alpha}}_i$) for $n = 1000$ and $T = 1024$

Tables 4.2 and 4.3 give an impression, how often for the 1000 generations we have the case of choosing an unsuitable model and which model is chosen then.

		d_{m_2}					
		0	**1**	**2**	**3**	**4**	**5**
	0	0	0	0	0	0	0
	1	0	2	0	0	0	0
d_{m_1}	**2**	0	0	0	0	0	0
	3	0	0	0	0	0	0
	4	0	0	0	0	0	0
	5	0	998	0	0	0	0

Table 4.2: Frequency of selection of a given model from 1000 generations of a TVAR(2) process of type TRIG1

		d_{m_2}				
		0	**1**	**2**	**3**	**4**
	0	0	1	0	0	0
	1	45	0	0	0	0
d_{m_1}	**2**	0	0	0	0	0
	3	0	933	3	0	0
	4	0	18	0	0	0

Table 4.3: Frequency of selection of a given model from 1000 generations of a TVAR(2) process of type TRIG2

From the (real) TVAR process that was used it is clear that the true order of the second coefficient is $d_{m_2}=1$. Considering the estimation of $a_1(u)$ there is no true order except for the case of the curve TRIG1. Figure 4.5 accordingly allows a graphical comparison between the simulated and the mean of the estimated coefficient functions.

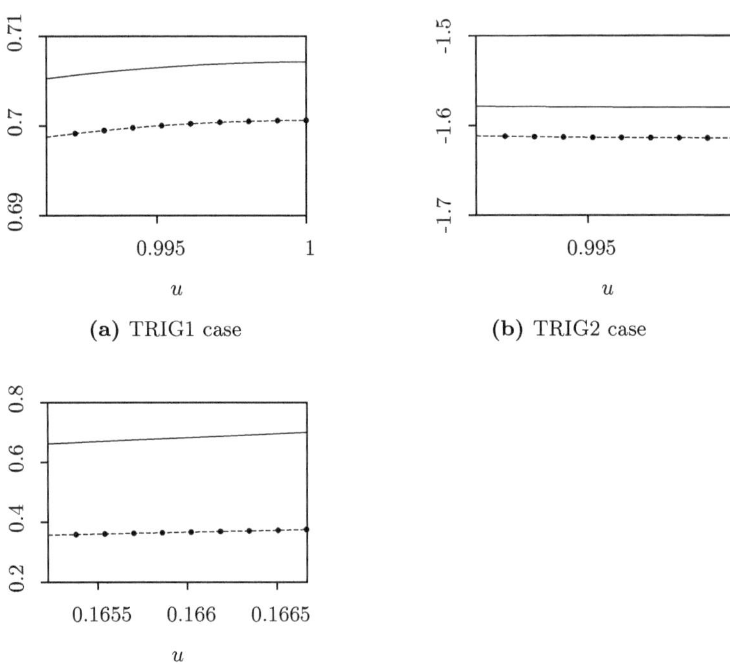

(a) TRIG1 case

(b) TRIG2 case

(c) LEG case

Figure 4.6: Coefficient extrapolation (— true curve α_1, -•- mean estimated curve $\bar{\hat{\alpha}}_1$) and $T = 1024$

Figure 4.6 shows the results for the coefficient extrapolations of Approach 1. It shows that the course of the extrapolated values resembles the true curve, especially for the TRIG 1 case. For the LEG case however, although the two curves are almost parallel, the difference between the true and the extrapolated values is very large. This shows that we did not manage to find a suitable ζ for the LEG case.

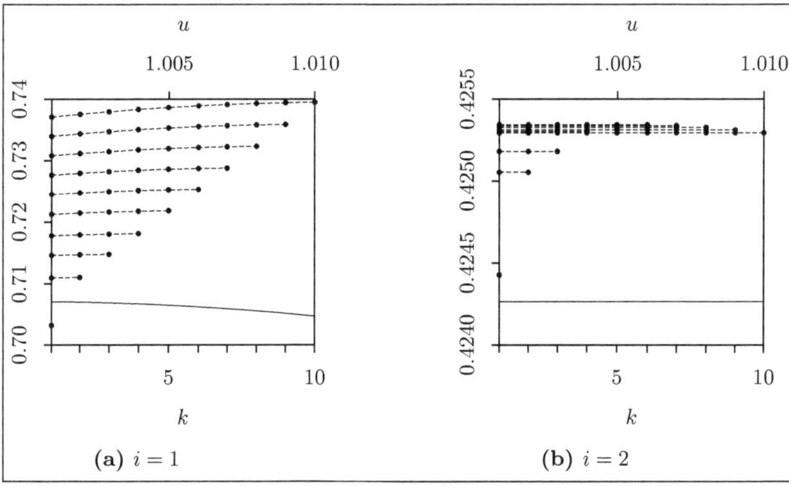

Figure 4.7: Time-dependent coefficient estimations based on APP3 and the penalized likelihood method (— true curve a_i, -•- mean estimated curve $\hat{\hat{a}}_i$) for the TRIG1 case and $T = 1024$

Yet, as we can see in Figure 4.7, using Approach 3 we get an additional problem. Using the same ζ for each of the extrapolation steps, the exrapolated values (in the case of a time-dependent coefficient) more and more depart from the true values. This could only be prevented if one would use different ζ for every extrapolation step. This could be a topic for future work on this field. But Approach 3 is not examined in this work any further.

Table 4.4 and 4.5 give the results for the mean square prediction errors, showing that the effects following from a different T are much bigger than those from different approaches. However, Approach 1 is superior to Approach 2 in most of our cases.

sample size		$T = 64$	
(true) polynomials	TRIG1	TRIG2	LEG
Approach 1	0.502	3.241	0.525
Approach 2	0.557	3.314	0.454
sample size		$T = 256$	
(true) polynomials	TRIG1	TRIG2	LEG
Approach 1	0.616	3.013	0.597
Approach 2	0.637	2.985	0.606
sample size		$T = 1024$	
(true) polynomials	TRIG1	TRIG2	LEG
Approach 1	0.458	0.503	0.610
Approach 2	0.451	0.554	0.619

Table 4.4: Median mean square errors for one-step ahead predictions from 1000 generations of a TVAR(2) process

sample size		$T = 64$	
(true) polynomials	TRIG1	TRIG2	LEG
Approach 1	0.665	3.455	0.464
Approach 2	0.715	3.703	0.616
sample size		$T = 256$	
(true) polynomials	TRIG1	TRIG2	LEG
Approach 1	0.636	4.454	0.648
Approach 2	0.763	4.806	0.655
sample size		$T = 1024$	
(true) polynomials	TRIG1	TRIG2	LEG
Approach 1	0.762	3.798	0.660
Approach 2	0.709	3.644	0.710

Table 4.5: Median mean square errors for three-step ahead predictions from 1000 generations of a TVAR(2) process

The large influence of the bandwidth b_T is illustrated in Figures A.2 and A.3. Our results here are contrary to the (theoretical) findings of Van Bellegem and von Sachs (2004) that an appropriate bandwidth is given (asymptotically) by $\log_2^{-3} T$. Also their recommendation to choose $b_T = \lceil \log_2 T \rceil = 10$ in practice clearly does not result in satisfying approximation results (even not for the second constant coefficient).

Against it, our results show that a good approximation goes along with a maximized bandwidth. That is why (except for the results in Figures A.2 and A.3), in this work, we always use $b_T = T$. However, the bandwidth selection should be a topic of further research as it quite has a great impact on the model size selection.

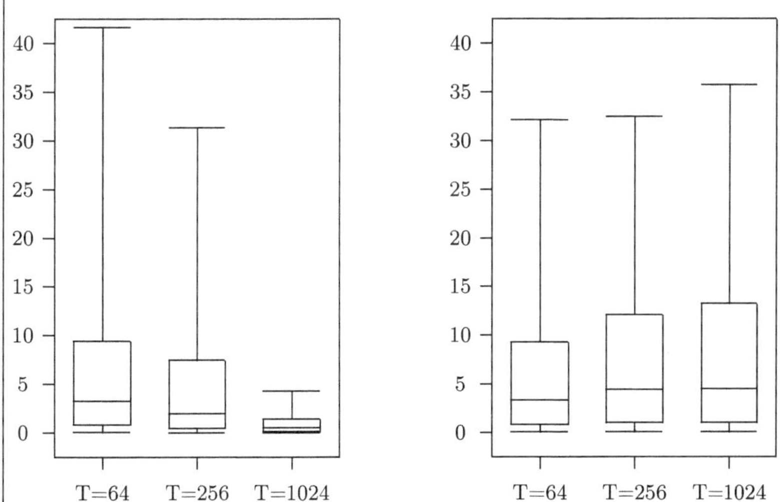

(a) TVAR model, TRIG2 case **(b)** ARIMA model, TRIG2 case

Figure 4.8: Mean square errors for one-step-ahead predictions (APP1) from 1000 generations of a TVAR(2) process; instead of the minimum and maximum values the 5 and 95 percent quantiles are shown

To answer the question, if using time-varying constant functions is advantageous compared to the use of constant coefficients, we draw a comparison to ARMA-models. Figure 4.8 shows that using TVAR processes we can profit from long sample sizes T, whereas with ARMA-modelling this is not the case.

And finally, in Figures A.4 and A.5, Approaches 1 and 2 are compared by the mean errors for ten-step-ahead predictions from 1000 generations of a TVAR(2) process with polynomials of the form TRIG1 and TRIG2. Whereas, for the TRIG2 case (Figure A.5) the differences between the two approaches are rather small, for the TRIG1 case (Figure A.4) Approach 1 leads to smaller errors that are also more stable.

5

Application

5.1 Motivation

As the above results show that in certain situations and for large sample sizes the proposed approach performs better in providing short-term-forecasts than the established method it is compared to, we wish to apply it to time series that are available for large sample sizes. As this is the case for many financial time series and as the research into price forecasting is of great interest, a reasonable next step is the practical evaluation of the proposed procedures by applying it to financial time series.

Of course, accurate price forecasts cannot be expected. The question here is rather as to whether the best forecast of tomorrow's price is simply today's price plus an estimate of the expected market interest rate, or if it is possible to draw profits from buying and selling the same goods many times over. So the question is, do two of the most important hypotheses for price research, the *random walk hypothesis* and the *efficient market hypothesis*, hold. Under the assumption of the random walk hypothesis prices wander in an entirely unpredictable way. And according to Fama (1976), a market is *perfectly efficient* if prices fully reflect available information, i. e. prices adjust fully when new information becomes available.

However, naturally neither the random walk hypothesis nor the efficient market hypothesis has been proven to hold. Although quite a lot work has been carried out in this field, there is by no means clarity concerning the adequacy of random walk processes for financial time series or the efficient market hypothesis. Taylor has investigated different surveys on stocks, commodity futures and currency exchange rates (see Taylor 1986 for a list of the most interesting books and articles) and states that the statistical methods for random walk tests are often inappropriate and that the methodology frequently used to assess trading rules is inadequate. From his investigations on different futures series he concludes that the efficient market hypothesis is probably false for several futures markets. That is why we will now focus on the analysis of futures series.

5.2 Futures data

The data that we analyse consist of daily exercise prices from the 27th March 2003 to the 22nd March 2007 of the *Euro-Bund-Future*,[1] which is a futures contract on a fictitious treasury bond of the Federal Republic of Germany with a duration of ten years and a coupon of six percent. The nominal amount is 100.000 Euros. The Euro-Bund-Future is listed at the *Eurex*, a stock exchange for time bargains, where it is possible to buy contracts for the next three dates of delivery. The date of delivery is the tenth calender day of the last month in every quarter.
As explained in the last section, it seems necessary for our approach to use long futures series. In our simulations in Section 4.4 we have achieved good results with series of length $T = 1024$. Thus it would be appropriate to have series with length of about 1000 prices. However, what we have is seventeen small time series each of them not longer then $T = 120$. Long series of futures prices can only be obtained if several contracts are used. But as we can see in Figure 5.1, the prices for different contracts differ. Another problem is that the prices in our dataset have not all been recorded at the same time of the day for different reasons. As we cannot compare prices recorded at completely different times, we cannot use the data that were recorded earlier than the others and have to skip them. We will explain in the next section how to construct one long time series from the time series for the different contracts and how to deal with the problem of missing values.

[1] The data have been purchased from the Karlsruher Kapitalmarktdatenbank (KKMDB).

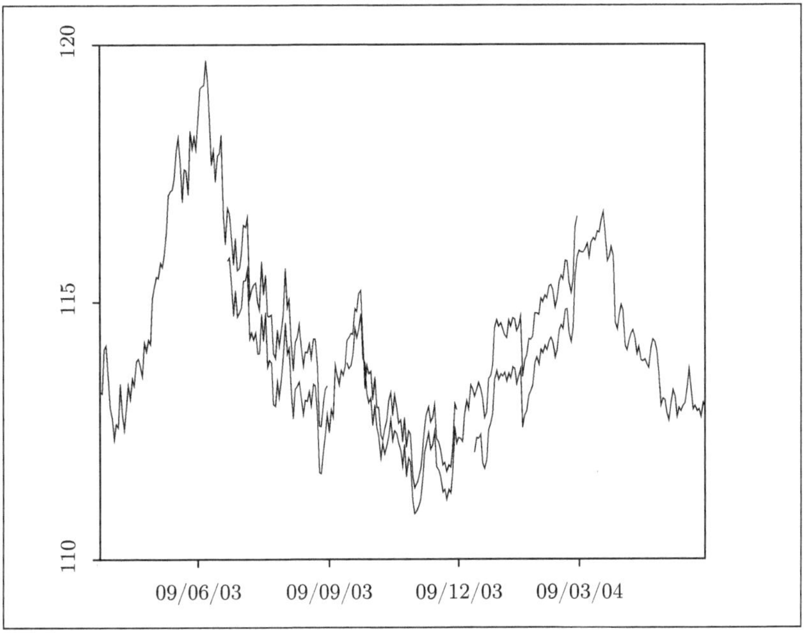

Figure 5.1: Daily exercise prices for the Euro-Bund-Future (in Euros) between March 2003 and June 2004 (expiration in September 2003, December 2003, March 2004, and June 2004)

5.2.1 Course of action

Let x_t denote the daily exercise price at time t, then we can obtain the (log-)d-days-return at time t by computing

$$r_t^d = \log \frac{x_t}{x_{t-d}} = \log x_t - \log x_{t-d} \,.$$

We make use of the log-returns as this makes computing different returns more convenient. Instead of computing every single series of returns, we just have to calculate the 1-day-returns and can then compute the other returns recursively using

$$
\begin{aligned}
r_t^d &= \log x_t - \log x_{t-d} \\
&= \log x_t - \log x_{t-1} + \log x_{t-1} - \log x_{t-2} + \ldots + \log x_{t-d+1} - \log x_{t-d} \\
&= r_t^1 + r_{t-1}^1 + \ldots + r_{t-d+1}^1 \\
&= \sum_{\nu=1}^{d} r_{t-\nu+1}^1 .
\end{aligned}
$$

The log-returns have to be calculated for each of the seventeen short time series belonging to a certain contract. As described above, we have to solve the problem of missing values before we can do that. Noack and Schlittgen (2000) have found in simulation studies that in most cases least squares estimations perform very well in estimating missing values in time series. We follow their recommendation. Then we construct one long time series by always using the returns of the series whose contract will expire next. As we can see in Figure 5.1 time series for different contracts do have similar trends, but they have different levels, which is due to the different residual time until their expiration. To avoid leaps in the constructed time series we have to adjust every single part belonging to a certain contract to the one thereafter. Let us suppose that we have three contracts. Contract i expires at time t^i, contract j at time t^j, and contract k at time t^k. Further let $t^i < t^j < t^k$ and contract k be the current contract. We define

$$\Delta_j := x_{t^j}^k - x_{t^j}^j ,$$

where x_t^k is the price for contract k and x_t^j the price for contract j at time t. We obtain the adjusted x_t^j for $t \leq t^j$ from

$$\tilde{x}_t^j = x_t^j + \Delta_j \frac{t - t^i}{t^j - t^i} .$$

We have to execute this adjustment for every transition from one contract to the following starting with the last one (March 2007) and going on until the first one (June 2003). We then obtain the log-returns for the whole time span. The log-returns and 50-days-log-returns are shown in Figure 5.2a and 5.2b respectively.

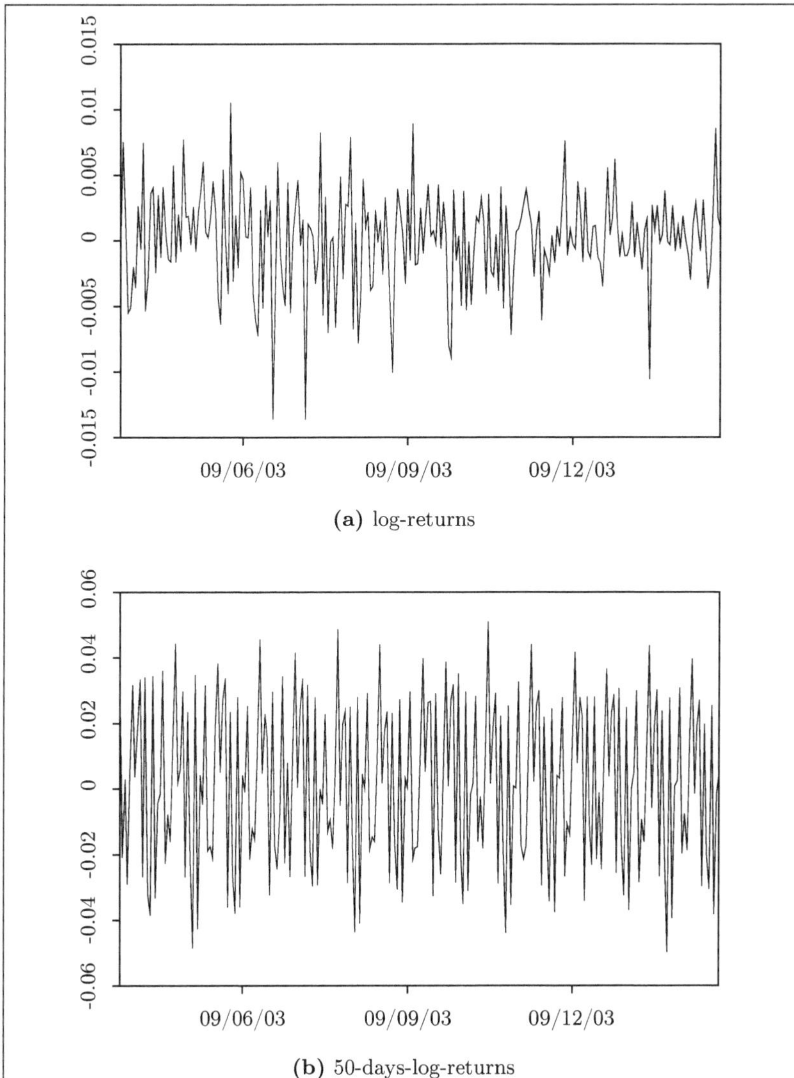

Figure 5.2: Log-returns and 50-days-log-returns for the Euro-Bund-Future (calculated from daily exercise prices) from March 2003 to March 2004

5.2.2 Practical evaluation of TVAR processes on futures series

To illustrate the time dependence of the AR-coefficients and hence the covariance instationarity we estimate the coefficients of an AR(4) process for 50-day sections of the log-returns, i. e. we calculate the coefficients at times $t = 1, \dots, 50$, repeat this for $t = 2, \dots, 51$, and so forth. The proceeding is illustrated in Figure 5.3. The smoothed results for the first coefficient can be found in Figure 5.4. As expected from the considerations in Section 5.1 the coefficients do not seem to be constant which makes the use of methods for stationary processes inadequate.

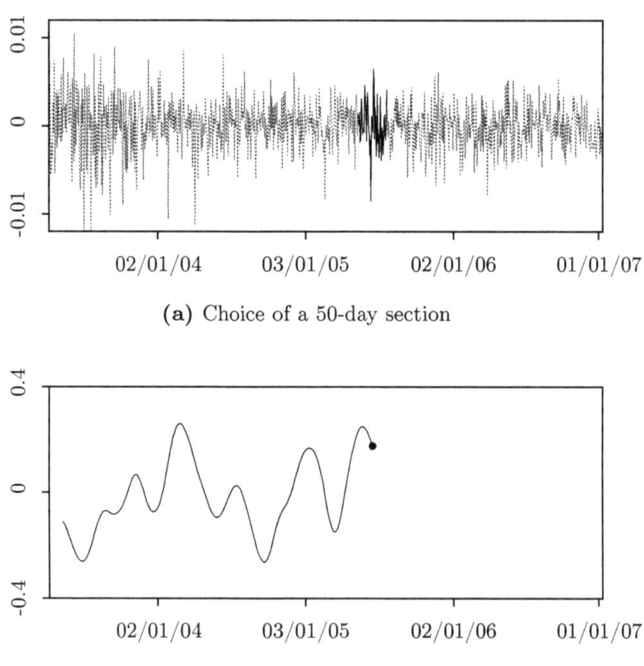

(a) Choice of a 50-day section

(b) Smoothed coefficient estimations for the chosen 50-day section (•) and behaviour of the coefficient estimations for previous sections (—)

Figure 5.3: Estimation procedure for the coefficients in Figure 5.4

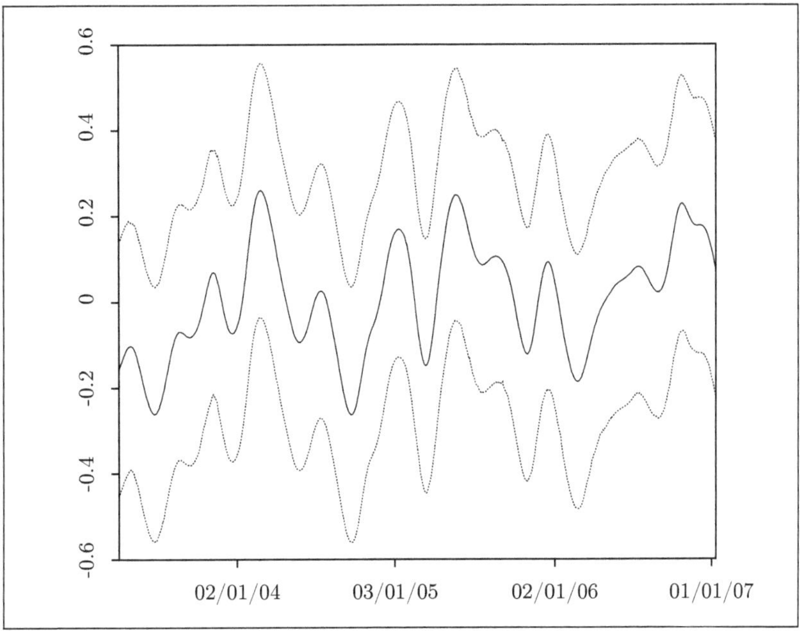

Figure 5.4: Smoothed coefficient estimations (parameter α_1) of AR(4) processes for the log-returns

We now calculate forecasts for these returns by modulating a TVAR process as described in Section 3.2. For the choice of the model size, i. e. the parameter p (or an adequate subset model) we have to compute the local partial autocorrelations as explained in Section 3.2. The results are shown in Figure 5.5. Obviously, the local partial autocorrelations of lags 4, 8, and 12 are significantly different from zero. That is why we choose a TVAR(4) process. As we want to obey the order of parsimony, we also examine different subset models. Thereby, we follow the iterative proceeding that is explained in Section 4.3 choosing the most parsimonious model, for which the residuals resemble a white noise process, i. e. the local autocorrelation and the local partial autocorrelation functions do not show values that are obviously significantly unequal from zero.

Finally, this leads us to the choice of the model

$$X_{t,T} + \alpha_1\left(\frac{t}{T}\right) X_{t-1,T} + \alpha_2\left(\frac{t}{T}\right) X_{t-2,T} + \alpha_4\left(\frac{t}{T}\right) X_{t-4,T} = e_{t,T}. \quad (5.5)$$

After fitting the model from Equation (5.5) to the data, we can perform predictions using Approach 1 from Section 4.2.

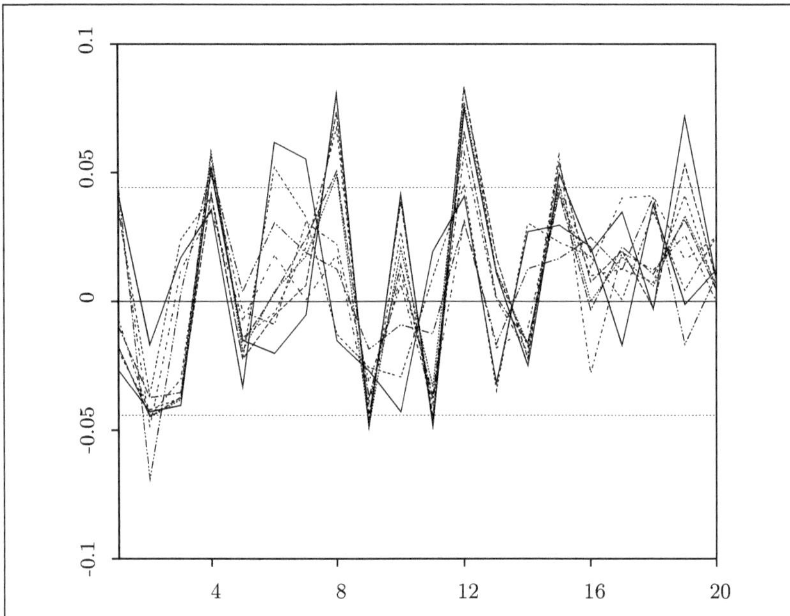

Figure 5.5: Local partial autocorrelations for the log-returns

Figure 5.6 shows the local partial autocorrelations for the 50-days-log-returns. Obviously, for these log-returns there are significant values for several lags, especially for lag 17. Fitting a TVAR(17) model to our data is very computationally intensive. Fitting more parsimonious subset models to the data (with two or three coefficient functions) results in residuals that are obviously not realizations of a white noise process. For this reasons we will only examine the (one-day-)log-returns from now on.

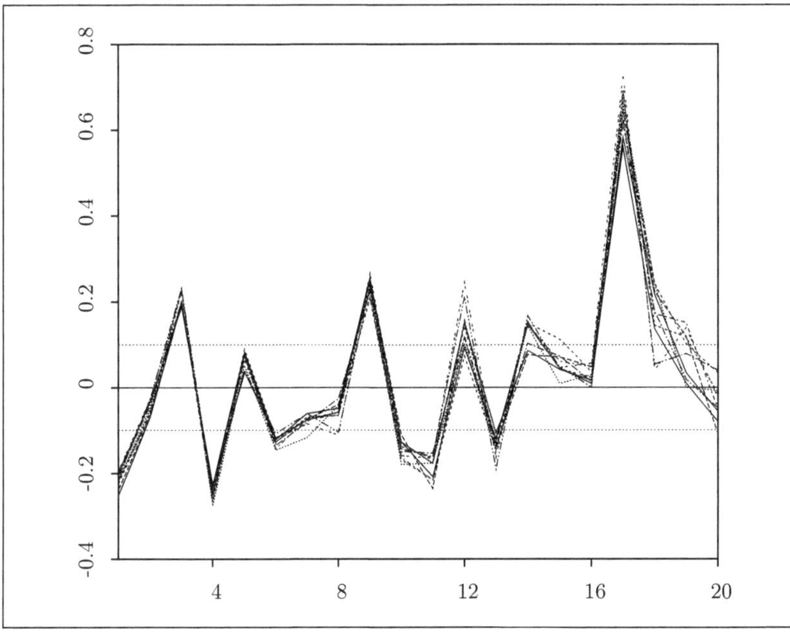

Figure 5.6: Local partial autocorrelations for the 50-days-log-returns

As we wish to measure the accuracy of our forecasts we skip the last twenty observations of the time series and predict these twenty values by using the remainder observations. We use *rolling forecasts*, i. e. after that we skip the last 21 values and try to predict the first twenty of them by using the remainder observations et cetera. We do this until we have performed twenty 1-to-20-step-forecasts for the log-returns. As we wish the time series not to differ in length we also skip values at the beginning. We compute the root mean square errors (RMSEs) of these forecasts for the different forecast horizons h. As a benchmark we also compute the RMSEs for forecasts resulting from exponential smoothing.[2]

[2] Another possible benchmark is given by so called *adaptive exponential smoothing* (Günther 1980). In this approach, the smoothing parameter varies depending on the recent forecasting accuracy. However, the findings of Ekern 1981 do not confirm that adaptive exponential smoothing methods provide superior forecasts compared to those obtained from constant exponential smoothing. For this reason only constant exponential smoothing is used in this work.

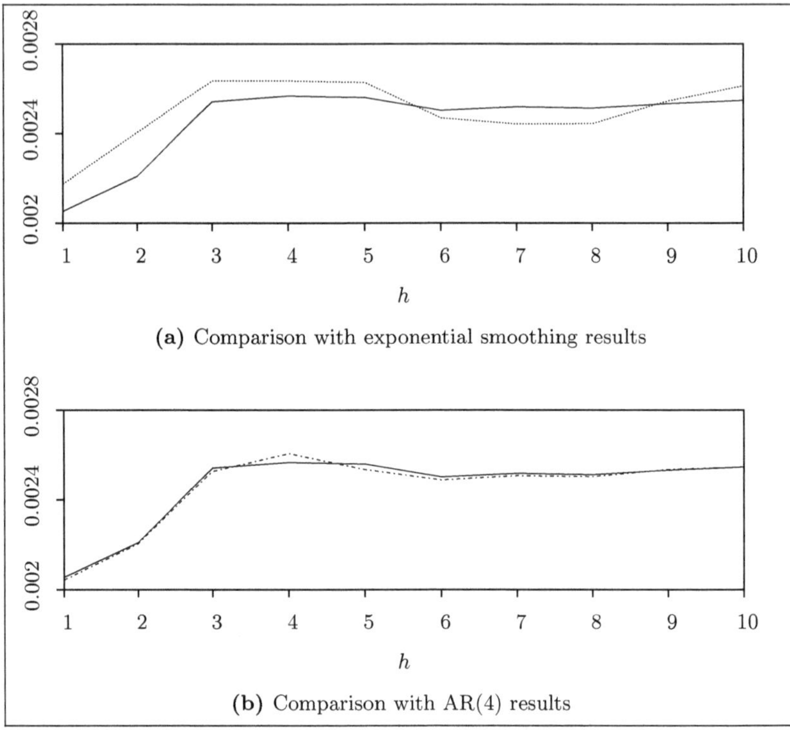

Figure 5.7: RMSEs for forecasts using exponential smoothing (···), AR(4)-processes for 50-day-sections (- · -) and Approach 1 from Section 4.2 (—) with $\zeta = 0.8$

The results for the log-returns can be found in Figure 5.7a. These show that we can obviously profit from our approach compared to the method of exponential smoothing for short forecasting horizons. Of course, admittedly the method of exponential smoothing is very naive. As our coefficient estimations in Figure 5.4 differ from zero, as a second benchmark, we compute forecasts using only the last fifty observations and modulating an AR(4) process for these observations (thereby we use the same subset as for the TVAR(4) processes). The results for the corresponding RMSEs are illustrated in Figure 5.7b.

They show that, for our data, some kind of local AR(4) process seems to be adequate. The two approaches result in similar forecasting errors.

5.3 Dow Jones index data

As a second example we want to apply the TVAR model to stock returns since, as explained in the introduction, in recent years it has been observed that these data often show time-varying second-order characteristics. Van Bellegem and von Sachs (2002) study the question of covariance stationarity for the Nasdaq index and conclude that it shows non-stationary behaviour. Anderson (1977, pp. 107 et seq.) modulates an AR process to the data of the Dow Jones utilities index, examining a data set from the early seventies. Different from the current data, in those days the estimations for the partial autocorrelations obviously show that, for small lags, these are significantly different from zero. So for these data there are obviously autocorrelations. Here, it remains the question, if we can compute better forecasts, if we assume that the real parameters are not constant, but that they are time-dependent.

The dataset that we analyse consists of the log-returns from February 2nd 1970 to October 19th 1972 of the Dow Jones Utility Average (also known as the *Dow Jones Utilities*, calculated from daily closing prices, 690 observations),[3] which is a stock index that keeps track of the performance of 15 prominent utility companies. It was created in 1929 when all utility stocks were removed from the Dow Jones Industrial Average. The log-returns can be found in Figure 1.2 and the daily closing prices in Figure 1.1.
In order to get a first impression if the coefficients are time-dependent, we follow the proceeding from above and compute the coefficient estimations for 50-day-sections of AR(2) processes. The smoothed results (Figure 1.3) for the first coefficient show the time dependence. It is remarkable that this time, different to the futures example, the estimations do scatter around zero, but are clearly positive for the whole sample.

[3] This dataset has been purchased from Yahoo!Finance (finance.yahoo.com).

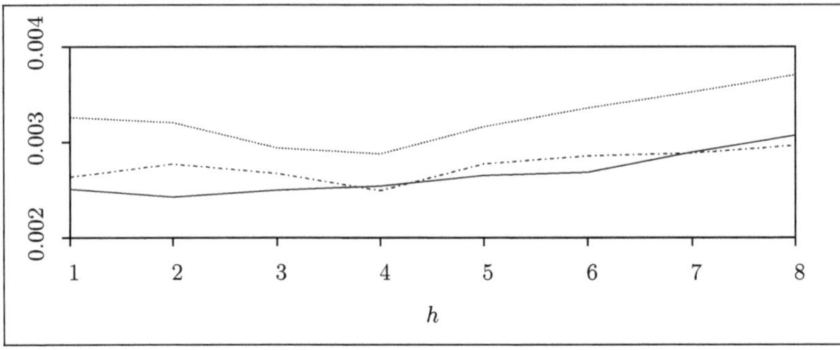

Figure 5.8: RMSEs for forecasts using exponential smoothing (···), AR(2)-processes for 50-day-sections (- · -) and Approach 1 from Section 4.2 (—) with $\zeta = 0.3$

Again we follow the same estimation and forecasting procedures as in the above futures example, which leads us to a TVAR(2) process this time. We achieve the results in Figure 5.8. Obviously, for this dataset the advantage of our model compared to the two benchmark models is still larger than in the futures example.

6

Conclusion

6.1 Contributions

The contributions of this thesis are of four types. Besides enhancements concerning the *model selection* and the *estimation* procedure we propose a *forecasting* approach for TVAR processes and provide *applications* of the proposed procedures to financial time series.

Model selection

In the semiparametric fitting we assume that the number of components of $\boldsymbol{\theta}$ is unknown. In the context of time-varying AR(p) fitting, this means that the order p is unknown. A procedure for selecting this order from data is proposed utilizing a local version of the partial autocorrelation function.

Estimation

The parameter functions are estimated by a semiparametric method. As the true coefficient functions are usually unknown, we have to approximate them. Therefore, in Chapter 4.2, we have provided a set of approximating functions.

Forecasting

Furthermore, in Chapter 4.2, we develop a method for adjusting the coefficient functions into the future and compare our approach with classical forecasting methods like Box-Jenkins-ARIMA-modelling using Monte

Carlo simulations. Thereby, the effects of varying sample sizes, diverse coefficient functions, and different approaches to predict the (future) coefficients are examined. If sample sizes are large and forecasting horizons do not range too far into the future, our approach turns out to be superior to classical methods. This is due to the good approximation of the coefficient functions.

Application

Finally, in Chapter 5 a practical evaluation of the proposed procedure is given by applying it to the Dow Jones Utility index and to futures prices.

6.2 Possible directions for future research

Some problems remain for future research.

Modelling

This work focuses on TVAR processes. A natural next step would be to investigate the more general class of TVARMA processes. Besides only models with stationary innovation processes are examined. A more realistic ansatz is to assume that the innovation processes are non-stationary. One interesting possibility would be to use GARCH processes (see Bollerslev 1982). A recursive algorithm for estimating time-varying ARCH processes (see Engle 1982) has already been given by Dahlhaus and Subba Rao (2007).

Model selection

The choice of a convenient smoothing method and the bandwidth selection should be a topic of further research as it quite has a great impact on the model size selection. Also a formal proof showing the asymptotical distribution of the local partial autocorrelation estimator should be derived.

Estimation

The selection of the factor ζ has to be investigated in more details.

Forecasting

Concerning the forecasting approach, this thesis does not provide an approach for computing prediction intervals. Furthermore, our results depend on simulations. So a next step would be the provision of theoretical results concerning goodness-of-fit of the predictions.

References

Anderson, O. (1977), *Time series analysis and forecasting: the Box-Jenkins approach*, Butterworths, London.

Barron, A., Birgé, L., and Massart, P. (1999), Risk bounds for model selection via penalization, *Probability Theory and Related Fields*, vol. 113, no. 3, pp. 301–413.

Birgé, L. and Massart, P. (1998), Minimum contrast estimators on sieves: exponential bounds and rates of convergence, *Bernoulli*, vol. 4, no. 3, pp. 329–375.

Bollerslev, T. (1982), Generalized autoregressive conditional heteroskedasticity, *Econometrica*, vol. 50, pp. 987 –1008.

Box, G. and Jenkins, G. (1976), *Time series analysis: forecasting and control*, Holden-Day, San Francisco.

Brillinger, D. (1981), *Time series: data analysis and theory*, Holden-Day, San Francisco.

Brockwell, P. J. and Davis, R. A. (1991), *Time Series: Theory and Methods*, 2nd ed., Springer, New York.

Comte, F. (2001), Adaptive estimation of the spectrum of a stationary Gaussian sequence, *Bernoulli*, vol. 7, no. 2, pp. 267–298.

Dahlhaus, R. (1996), On the Kullback-Leibler information divergence of locally stationary processes, *Stochastic Processes and their Applications*, vol. 62, no. 1, pp. 139–168.

— (1997), Fitting time series models to nonstationary processes, *The Annals of Statistics*, vol. 25, no. 1, pp. 1–37.

— (2000), A likelihood approximation for locally stationary processes, *The Annals of Statistics*, vol. 28, no. 6, pp. 1762–1794.

— (2009), Local inference for locally stationary time series based on the empirical spectral measure, *Journal of Econometrics*, vol. 151, pp. 101–112.

Dahlhaus, R. and Subba Rao, S. (2007), A recursive online algorithm for the estimation of time-varying ARCH parameters, *Bernoulli*, vol. 13, pp. 389–422.

Dhrymes, P. J. (1974), *Econometrics: Statistical Foundations and Applications*, Springer, New York.

Doob, J. L. (1953), *Stochastic processes*, Wiley, New York.

Ekern, S. (1981), Adaptive Exponential Smoothing Revisited, *The Journal of the Operational Research Society*, vol. 32, no. 9, pp. 775–782.

Engle, R. F. (1982), Autoregressive Conditional Heteroscedasticity with Estimates of the Variance of United Kingdom Inflation, *Econometrica*, vol. 50, no. 4, pp. 987–1007.

Fama, E. F. (1976), *Foundations of finance: portfolio decisions and securities prices*, Basic Books, New York.

Fuller, W. A. (1996), *Introduction to Statistical Time Series*, Wiley, New York.

Geman, S. and Hwang, C.-R. (1982), Nonparametric maximum likelihood estimation by the method of sieves, *The Annals Of Statistics*, vol. 10, pp. 401–414.

Grenander, U. (1981), *Abstract inference*, Wiley, New York.

Günther, R. (1980), Adaptive exponentielle Glättung erster Ordnung, *Series Statistics*, vol. 11, pp. 327–340.

Hannan, E. (1970), *Multiple time series*, Wiley, New York.

Iacus, S. M. (2008), *Simulation and Inference for Stochastic Differential Equations With R Examples*, Springer, New York.

Katznelson, Y. (2004), *An introduction to harmonic analysis*, Cambridge University Press, Cambridge.

Kreiß, J. and Neuhaus, G. (2006), *Einführung in die Zeitreihenanalyse*, Springer, Heidelberg,

Kullback, S. (1959), *Information theory and statistics*, Wiley, New York.

Kullback, S. and Leibler, R. A. (1951), On information and sufficiency, *The Annals Of Mathemathical Statistics*, vol. 22, no. 1, pp. 79–86.

Loretan, M. and Philips, P. C. B. (1994), Testing the covariance stationarity of heavy-tailed time series, *Journal of Empirical Finance*, vol. 1, pp. 211–248.

Meintrup, D. and Schäffler, S. (2005), *Stochastik: Theorie und Anwendungen*, Springer, Berlin.

Neumann, M. H. and von Sachs, R. (1997), Wavelet thresholding in anisotropic function classes and application to adaptive estimation of evolutionary spectra, *The Annals of Statistics*, vol. 25, no. 1, pp. 38–76.

Noack, T. and Schlittgen, R. (2000), Nonparametric estimation of missing values in time series, *Allgemeines Statistisches Archiv*, vol. 84, pp. 23–32.

Pagan, A. R. and Schwert, G. W. (1990), Testing for covariance stationarity in stock market data, *Economics Letters*, vol. 33, pp. 165–170.

Parzen, E. (1983), in: Elsevier, Amsterdam, chap. 11, Autoregressive spectral estimation, pp. 221–247.

Pourahmadi, M. (2001), *Foundations of time series analysis and prediction theory*, Wiley, New York.

Priestley, M. B. (1965), Evolutionary Spectra and Non-Stationary Processes, *Journal of the Royal Statistical Society: Series B (Statistical Methodology)*, vol. 27, no. 2, pp. 204–237.

— (1981), *Spectral analysis and time series*, Academic Press, London.

Rudin, W. (1987), *Real and complex analysis*, 3rd ed., McGraw-Hill, New York.

Schlittgen, R. and Streitberg, B. H. J. (2001), *Zeitreihenanalyse*, 9th ed., Oldenbourg Wissenschaftsverlag, München.

Serfling, R. (2002), *Approximation theorems of mathematical statistics*, Wiley, New York.

Shannon, C. E. (1948), Mathematical Theory of Communication, *The Bell System Technical Journal*, vol. 27, no. 3, pp. 379–423.

Taylor, S. (1986), *Modelling Financial Time Series*, Wiley, Chichester.

Tong, H. (1993), *Non-linear time series: a dynamical system approach*, Clarendon Press, Oxford.

Van Bellegem (2003), Adaptive Methods for modelling, estimating and forecasting locally stationary processes, PhD thesis, Université Catholique de Louvain, Louvain-la-Neuve.

Van Bellegem, S. and von Sachs, R. (2004), On adaptive estimation for locally stationary wavelet processes and its applications, *International Journal of Wavelets, Multiresolution and Information Processing*, vol. 2, pp. 545–565.

Van Bellegem, S. and Dahlhaus, R. (2006), Semiparametric estimation by model selection for locally stationary processes, *Journal of the Royal Statistical Society: Series B (Statistical Methodology)*, vol. 68, no. 5, pp. 721–746.

Van Bellegem, S. and von Sachs, R. (2002), Forecasting economic time series using models of nonstationarity, *Discussion Paper 0227*, Institut de Statistique, Université catholique de Louvain, Louvain-la-Neuve.

— (2008), Locally adaptive estimation of evolutionary wavelet spectra, *The Annals of Statistics*, vol. 36, no. 4, pp. 1879–1924.

Notations and abbreviations

List of symbols

$\ldots'$	transpose of a vector or matrix	32
$\overline{\cdots}$	complex conjugate	20
$\lesssim$	less or similar	52
$\otimes$	direct product	52
$\square$	end of proof	34
$\langle\cdot,\cdot\rangle$	inner product	21
$\lceil\cdot\rceil$	floor function	40
$\lvert\cdot\rvert$	modulus	20
$\lVert\cdot\rVert_{\text{spec}}$	spectral norm	55
$\lVert\cdot\rVert_2$	euclidian norm	21
$\xrightarrow{\mathcal{D}}$	convergence in distribution	41
$\{e_t\}$	white noise process	26
$\{x_t\}$	time series	20
$\{X_t\}$	stochastic process	20
$(\Omega, \mathcal{A}, \mathrm{P})$	probability space	20
$\mathcal{A}$	sigma algebra	20

$A^{\circ}(\cdot)$	transfer function	15
$A_t(\cdot)$	time-varying transfer function	14
b_T	bandwidth	40
$\mathbb{C}$	set of complex numbers	20
Cor	correlation	29
Cov	covariance	40
$d(\cdot\|\cdot)$	Kullback-Leibler information divergence	46
det	determinant	30
$d_{\boldsymbol{m}}$	dimension of a sieve	55
$\mathscr{D}(\cdot,\cdot)$	asymptotic Kullback-Leibler information divergence	49
$\mathrm{E}(\cdot)$	expectation	20
$f(\cdot)$	spectral density	23
$f_t(\cdot)$	evolutionary spectral density	35
$f_T(\cdot,\cdot)$	Wigner-Ville spectrum	39
$f_{\boldsymbol{\theta}}(\cdot)$	spectral density with semiparametric structure	54
$\mathcal{F}$	family of functions	15
$F(\cdot)$	spectral distribution	24
$\mathcal{F}^D$	finite-dimensional approximation space	51
$g(\cdot)$	density	46
$\mathcal{G}$	set of information	59
$G(\cdot)$	distribution	21
i	imaginary unit	23
I	periodogram	46
J_T	preperiodogram	50
$K(\cdot)$	kernel	40
l	likelihood	46

$\lim_{T\to\infty} E$	convergence in mean	48
log	natural logarithm	46
L	lag operator	27
$L^2_{\mathbb{C}}$	space of square integrable, complex valued random variables	20
$\mathcal{L}_T$	Whittle likelihood	48
$\mathcal{M}_T$	set indexing a collection of sieves	52
$\mathbb{N}$	set of natural numbers	22
o	small Landau set	39
$\mathcal{O}$	big Landau set	52
pen(·)	penalty function	55
P	probability measure	20
$\boldsymbol{P}_k$	$k \times k$ autocorrelation matrix	30
$\mathbb{R}$	set of real numbers	19
$R^2_{i.x}$	coefficient of determination for $\sigma_{ii.x}$	34
$\overline{sp}$	closed span	60
sup	supremum	37
t	time point	19
T	sample size	37
$\mathcal{T}$	set of time indices	19
TV	total variation norm	39
u	time point in rescaled time	15
Var	variance	33
x_t	observation at time t	19
X_t	random variable at time t	20
$\mathbb{Z}$	set of integers	23
$Z(\cdot)$	orthogonal increment process	14

α_j	jth coefficient of an AR(q) process	26
$\alpha_j(\cdot)$	jth coefficient function of a TVAR(p) process	16
β_j	jth coefficient of a MA(q) process	27
$\gamma(\cdot)$	autocovariance	20
$\delta(\cdot)$	Dirac-delta-function	37
$\boldsymbol{\theta}_\circ$	target parameter curve	48
$\boldsymbol{\theta}(\cdot)$	D-dimensional vector of coefficient functions	16
$\hat{\boldsymbol{\theta}}_{\boldsymbol{m}}$	minimum contrast estimator on a sieve $\mathcal{F}_{\boldsymbol{m}}$	52
λ	frequency	14
λ_k	Fourier frequency	46
μ	mean	20
σ	standard deviation	16
$\boldsymbol{\Sigma}$	covariance matrix	55
$\phi(\cdot)$	partial autocorrelation	29
Ω	sample space	20

List of abbreviations

ARMA(p, q)	autoregressive moving average process of orders p and q	28
AR(p)	autoregressive process of order p	13
i. i. d.	independent, identically distributed	36
MA(q)	moving average process of order q	27
TVAR(p)	time-varying autoregressive process of order p	16

List of tables

List of figures

A

Appendix

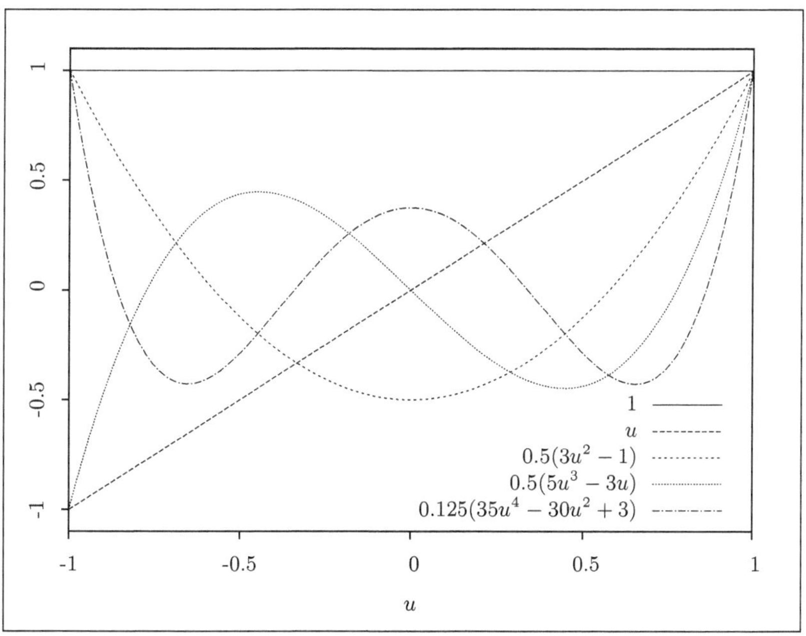

Figure A.1: First five Legendre polynomials

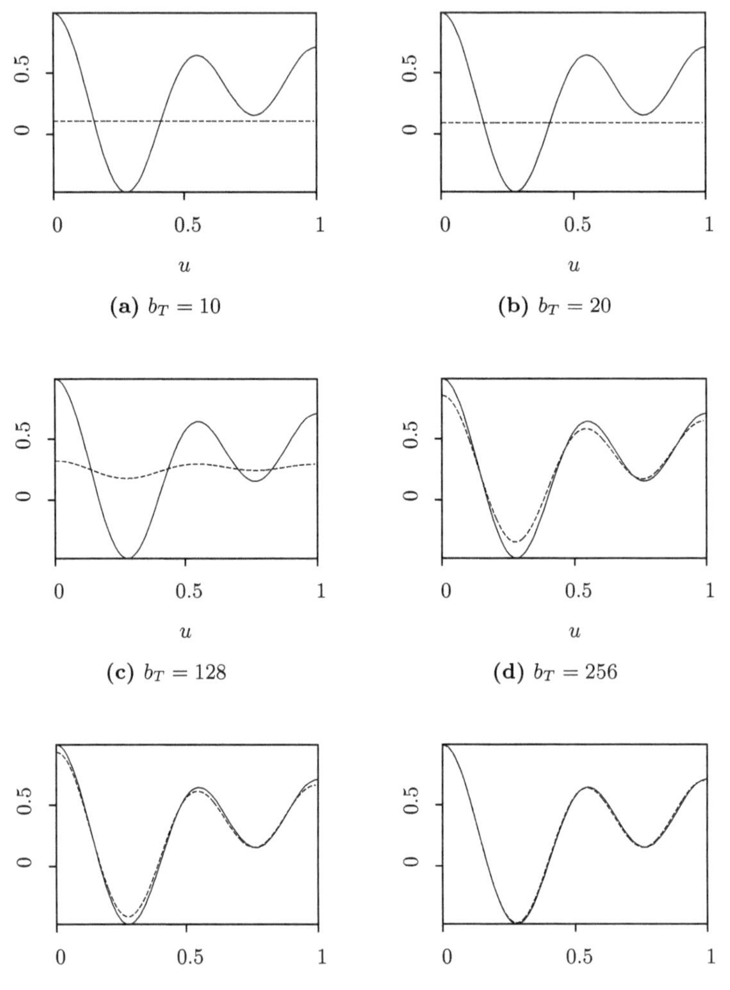

Figure A.2: Time-dependent coefficient estimations based on APP1 and the penalized likelihood method (— true curve a_i, - - - mean estimated curve $\bar{\hat{a}}_i$) for the TRIG1 case, $T = 1024$, and different bandwidths b_T

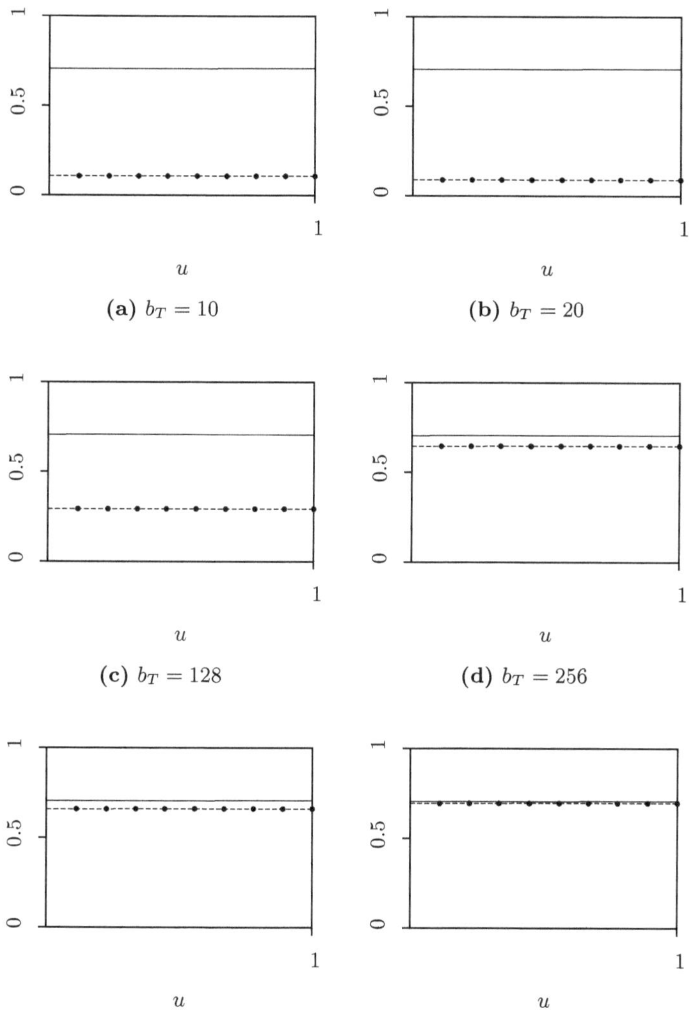

(a) $b_T = 10$ (b) $b_T = 20$ (c) $b_T = 128$ (d) $b_T = 256$ (e) $b_T = 512$ (f) $b_T = 1024$

Figure A.3: Time-dependent coefficient extrapolations based on APP1 and the penalized likelihood method (— true curve a_i, - - - mean estimated curve $\bar{\hat{a}}_i$) for the TRIG1 case, $T = 1024$, and different bandwidths b_T

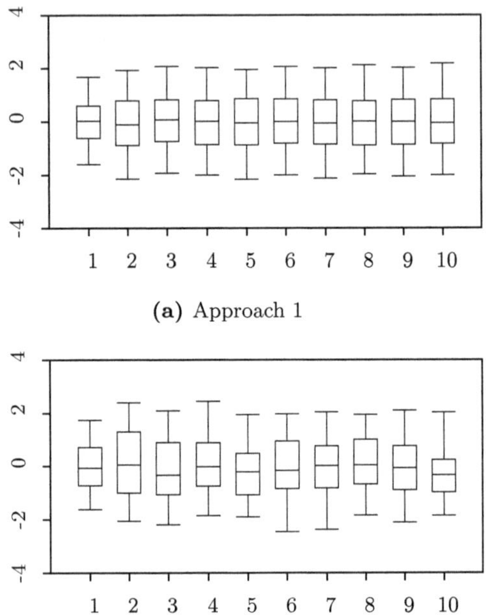

(a) Approach 1

(b) Approach 2

Figure A.4: Mean errors for ten-step-ahead predictions from 1000 generations of a TVAR(2) process with polynomials of the form TRIG1; the sample size is $T = 1024$; instead of the minimum and maximum values the 5 and 95 percent quantiles are shown

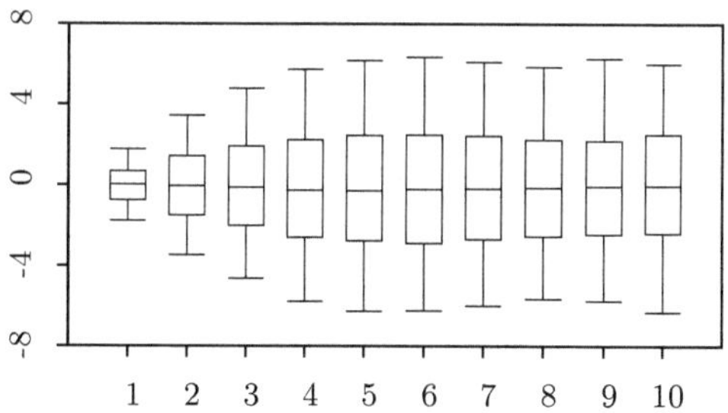

(a) Approach 1

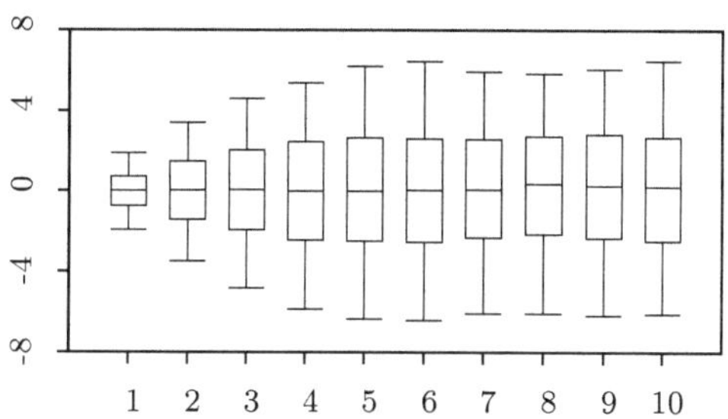

(b) Approach 2

Figure A.5: Mean errors for ten-step-ahead predictions from 1000 generations of a TVAR(2) process with polynomials of the form TRIG2; the sample size is $T = 1024$; instead of the minimum and maximum values the 5 and 95 percent quantiles are shown

B

GAUSS source code

B.1 Fitting time-varying autoregressive models to non-stationary processes

B.1.1 `modelselect()`

Purpose

Choosing the model with the minimum objective function

Input

`x` $(n, 1)$-data-vector
`sub` $(p, 1)$-vector, determines the subset model that is used; if an element is 0, it is expected that the coefficient for the respective lag is not significant; if 1, it is included
`dim` scalar, non-negative integer by which the dimension of the spaces of approximation is bounded from above
`M_T` number of off-diagonals of the preestimated covariance matrix which is used in the computation of the penalty function
`k` number of prediction steps
`pm` scalar, forecasting method:
  if `pm = 1`, coefficients computed for each specific time point;
  if `pm = 2`, constant coefficients;
  if `pm = 3`, recursive computation of forecasts
`print_cov` if 1, only the covariance matrix is printed;
  if 0, complete model selection
`m_min_chosen` $(\texttt{dim}, 1)$-vector of dimensions; if 0, the vector `m_min_chosen` will be chosen during the run of the procedure
`bandw` scalar, non-negative integer $\leq n$, bandwidth used for the smoothing step
`freq` quotient of observed time and observed time plus extra time until the coefficient function takes the initial value again

Output

`x_pred` predicted values
`x_fitted` fitted values
`x_resid` residuals
`parV` parameter vector that minimizes the objective function
`min_m` number of parameters for the minimal objective function
`au_ges` vector of estimated coefficients
`au_pred` vector of predicted coefficients

```
#include qnewtonmt.sdf
proc(7)=
modelselect(x,sub,dim,M_T,k,pm,print_cov,m_min_chosen,bandw,freq);
local
acfmat,au,au_ges,au_pred,bigGamma,bigSigma,bigSigma_spec,bigSigma_upper,bigX,C,
d,datarray,dim_smallC,dim3_dat,dim4_dat,Gamma_tilde,i,j,k4,loc,m,m_min_sumL,
mask,mask_ones,min_m,min_sumL,min_sumL0,n,n_mani,orders,p,pacfmat,pardim,ParN,
ParV,pcf,pdim,phi,phi_array,phi_tilde,pmat0,q,r,s,s_par,sigma,smallC,
smallC_smooth,smallC_smooth_der,step1,step2,t,theta,theta_str,v,w,x_fitted,
x_mani,x_pred,x_resid,zeroafter,zerobefore;
struct QNewtonmtControl c0;
  c0=QNewtonmtControlCreate;
  c0.output=0;
  c0.CovType=0;
  c0.MaxIters=100;
  c0.MaxTries=0;
  c0.relGradTol=1e-4;
  c0.randRadius=0;
  c0.Printiters=0;
struct DS d0;
  d0=dsCreate;
p=rows(sub);
if M_T<=p;
  dim_smallC=p+1;
else;
  dim_smallC=M_T+1;
endif;
zeroafter=zeros(ceil((dim_smallC-1)/2),1);
if dim_smallC>2;
  zerobefore=zeros(floor((dim_smallC-1)/2),1);
  x_mani=zerobefore|x|zeroafter;
else;
  x_mani=x|zeroafter;
endif;
n=rows(x);
n_mani=rows(x_mani);
bigX=arrayinit(dim_smallC|n|1,0);
r=0;
do while r<dim_smallC;
  r=r+1;
  bigX[r,.,.]=x_mani[r:n_mani-dim_smallC+r];
endo;
bigX=atranspose(bigX,2|1|3);
smallC=arrayinit(n|dim_smallC|1,0);
```

```
r=0;
do while r<dim_smallC;
  r=r+1;
  smallC[.,r,.]=
  amult(bigX[.,floor(ceil((dim_smallC)/2)+r/2),.],
  bigX[.,floor(ceil((dim_smallC)/2)-(r-2)/2),.]);
endo;
Gamma_tilde=arrayinit(n|p+1|p+1,0);
r=0;
do while r<p+1;
  r=r+1;
  s=0;
  do while s<p+1;
    s=s+1;
    Gamma_tilde[.,r,s]=smallC[.,abs(r-s)+1,.];
  endo;
endo;
bigGamma=Gamma_tilde[.,1:p,1:p];
C=Gamma_tilde[.,1,2:p+1];
d=seqa(1,1,dim)';
phi_tilde
  =sqrt(2)*cos(freq*2*pi*(d-1).*seqa(1,1,n)/(n+k));
phi_array=arrayinit(1|n|dim,1);
phi_array[1,.,.]=phi_tilde;
phi=atranspose(phi_array,2|3|1);
if p<=dim;
  dim3_dat=dim;
elseif p>dim;
  dim3_dat=p;
endif;
if M_T<p;
  dim4_dat=p;
elseif M_T>=p;
  dim4_dat=M_T+1;
endif;
datarray=arrayinit(5|n|dim3_dat|dim4_dat,0);
datarray[1,.,1:p,1:p]=bigGamma;
datarray[2,.,1,1:p]=C;
datarray[3,.,1:dim,1]=phi;
smallC=atranspose(smallC,3|1|2);
smallC=getMatrix(smallC,1);
step1=seqa(1,1,n);
step2=seqa(1,0.5,2*n-1);
smallC_smooth=zeros(2*n-1,dim_smallC);
q=0;
do while q<dim_smallC;
  q=q+1;
{smallC_smooth[.,q],smallC_smooth_der,bandw}
  =nw(step2,step1,smallC[.,q],bandw,&k_rect);
endo;
bigSigma=zeros(n,n);
s=0;
do while s<n;
```

```
  s=s+1;
  t=0;
  do while t<n;
    t=t+1;
    if abs(s-t)<=M_T;
      bigSigma[s,t]=smallC_smooth[(s+t-1),abs(s-t)+1];
    endif;
  endo;
endo;
if print_cov==1;
  acfmat=zeros(n-M_T,M_T+1);
  pacfmat=zeros(n-M_T,M_T);
  i=0;
  do while i<n-M_T;
         i=i+1;
    acfmat[i,.] = bigSigma[i,i:i+M_T];
    pcf = ldrec(bigSigma[i,i:i+M_T]');
    pacfmat[i,.] = pcf[.,M_T+2]';
  endo;
  "pacfmat";
  pacfmat[1,.];
  "acfmat";
  acfmat[1,.];
  end;
endif;
bigSigma_spec=maxc(eigh(bigSigma));
datarray[5,1,1,1]=bigSigma_spec;
d0.dataarray=datarray;
sigma=1;
theta_str="theta"$+ftocv(seqa(1,1,p),2,0);
theta=zeros(p,dim);
m=ones(1,p);
mask=zeros(p,dim);
mask_ones=ones(p,1);
mask[.,1]=mask_ones;
pmat0=zeros(p,1);
min_sumL0=1000000;
if sumc(m_min_chosen)/=0;
  goto short;
endif;
do while sumc(pmat0)<sumc(sub)*dim;
  pmat0=count(pmat0,dim,sub);
  m=pmat0';
  struct PV par0;
  par0=pvCreate;
  par0=pvPack(par0,sigma,"sigma");
  s_par=0;
  do while s_par<p;
    s_par=s_par+1;
    if m[s_par]==0;
      goto jump1;
    endif;
    mask[s_par,.]=zeros(1,dim);
```

```
    mask[s_par,1:m[s_par]]=ones(1,m[s_par]);
    par0=pvPackm(par0,theta[s_par,1:m[s_par]],
      theta_str[s_par],mask[s_par,1:m[s_par]]);
    jump1:
  endo;
  d0.dataarray[4,1,1,1:p]=m;
  d0.dataarray[4,2,1,1]=p;
  d0.dataarray[4,3,1,1]=dim;
  d0.dataarray[4,4,1,1]=M_T;
  d0.dataarray[4,5,1,1:p]=sub';
  struct QNewtonMTOut out0;
  out0=QNewtonMT(&infodiv,par0,d0,c0);
  min_sumL=out0.fct;
  if min_sumL<min_sumL0;
    min_sumL0=min_sumL;
    min_m=m;
    struct PV par1;
    par1=pvCreate;
    par1=par0;
    struct DS d1;
    d1=dsCreate;
    d1=d0;
  endif;
endo;
m_min_sumL=min_sumL0;
par0=par1;
d0=d1;
out0=QNewtonMT(&infodiv,par0,d0,c0);
ParN=pvGetParNames(par1);
ParV=pvGetParVector(out0.par);
pardim=rows(parV);
pdim=0;
ParV=ParV';
i=0;
do while i<p;
  i=i+1;
  v=min_m[i];
  w=sumr(min_m[1:i]);
  if v>0;
    theta[i,1:v]=parV[w-v+2:w+1];
  endif;
endo;
x_pred=zeros(n+k,1);
x_pred[1:n]=x;
au_pred=zeros(p,k);
j=0;
do while j<k;
  j=j+1;
  if pm==1;
    au=coeff_theta(freq*(n+j)/(n+k),theta,min_m);
  elseif pm==2;
    au=coeff_theta(freq*n/(n+k),theta,min_m);
  endif;
```

```
    au_ges=coeff_theta(freq*seqa(0,1,n)/(n+k),theta,min_m);
    i=0;
    do while i<p;
      i=i+1;
      x_pred[n+j]=x_pred[n+j]-au[i]*x_pred[rows(x_pred)-k+j-i];
    endo;
    au_pred[1:cols(au),j]=au';
endo;
x_fitted=zeros(n-p,1);
x_resid=zeros(n-p,1);
i=p;
do while i<n;
  i=i+1;
  j=0;
  do while j<p;
          j=j+1;
    x_fitted[i-p]=x_fitted[i-p]-au_ges[i,j]*x[i-j];
    x_resid[i-p]=x[i]-x_fitted[i-p];
  endo;
endo;
goto finish;
short:
m=m_min_chosen;
struct PV par01;
par01=pvCreate;
par01=pvPack(par01,sigma,"sigma");
s_par=0;
do while s_par<p;
  s_par=s_par+1;
  if m[s_par]==0;
    goto jump11;
  endif;
  mask[s_par,.]=zeros(1,dim);
  mask[s_par,1:m[s_par]]=ones(1,m[s_par]);
par01=pvPackm
  (par01,theta[s_par,1:m[s_par]],theta_str[s_par],mask[s_par,1:m[s_par]]);
  jump11:
endo;
d0.dataarray[4,1,1,1:p]=m;
d0.dataarray[4,2,1,1]=p;
d0.dataarray[4,3,1,1]=dim;
d0.dataarray[4,4,1,1]=M_T;
struct QNewtonMTOut out01;
out01=QNewtonMT(&infodiv,par01,d0,c0);
min_sumL=out01.fct;
if min_sumL<min_sumL0;
  min_sumL0=min_sumL;
  min_m=m;
  struct PV par11;
  par11=pvCreate;
  par11=par01;
  struct DS d11;
  d11=dsCreate;
```

```
  d11=d0;
endif;
m_min_sumL=min_sumL0;
par01=par11;
d0=d11;
out01=QNewtonMT(&infodiv,par01,d0,c0);
ParN=pvGetParNames(par11);
ParV=pvGetParVector(out01.par);
pardim=rows(parV);
pdim=0;
ParV=ParV';
i=0;
do while i<p;
  i=i+1;
  v=min_m[i];
  w=sumr(min_m[1:i]);
  if v>0;
    theta[i,1:v]=parV[w-v+2:w+1];
  endif;
endo;
x_pred=zeros(n+k,1);
x_pred[1:n]=x;
au_pred=zeros(p,k);
j=0;
do while j<k;
  j=j+1;
  if pm==1;
    au=coeff_theta(freq*(n+j)/(n+k),theta,min_m);
  elseif pm==2;
    au=coeff_theta(freq*n/(n+k),theta,min_m);
  endif;
au_ges=coeff_theta(freq*seqa(0,1,n)/(n+k),theta,min_m);
  i=0;
  do while i<p;
    i=i+1;
    x_pred[n+j]=x_pred[n+j]-au[i]*x_pred[rows(x_pred)-k+j-i];
  endo;
  au_pred[1:cols(au),j]=au';
endo;
x_fitted=zeros(n-p,1);
x_resid=zeros(n-p,1);
i=p;
do while i<n;
  i=i+1;
  j=0;
  do while j<p;
          j=j+1;
    x_fitted[i-p]=x_fitted[i-p]-au_ges[i,j]*x[i-j];
    x_resid[i-p]=x[i]-x_fitted[i-p];
  endo;
endo;
finish:
```

```
retp(x_pred,x_fitted,x_resid,parV,min_m,au_ges,au_pred);
endp;
```

B.1.2 modelselect3()

```
proc(7)=modelselect3(x,sub,dim,M_T,k,print_cov,min_m_chosen,bandw,freq);
local
au,au_ges,au_pred,i,min_m,min_m_ges,n,p,parV,parV_ges,x_fitted,x_ges,x_resid;
p=rows(sub);
min_m=zeros(1,p);
n=rows(x);
au_ges=arrayinit(k|p|n+k-1,0);
i=0;
parV_ges=zeros(k,dim*p+1);
min_m_ges=zeros(k,p);
x_ges=x;
do while i<k;
  i=i+1;
  {x_ges,x_fitted,x_resid,parV,min_m,au,au_pred}
    =modelselect(x_ges, sub, dim, M_T, k-i+1, 4, print_cov, min_m ,bandw,freq);
  parV_ges[i,1:cols(parV)]=parV;
  min_m_ges[i,1:cols(min_m)]=min_m;
  au_ges[i,.,1:n+i-1]=au';
endo;
retp(x_ges,x_fitted,x_resid,parV_ges,min_m_ges,au_ges,au_pred);
endp;
```

B.1.3 infodiv()

Purpose

Objective function computing the distance between an estimate of the (data) evolutionary spectral density and the (model) evolutionary spectral density

Input

`struct PV par0` structure of type PV
`struct DS d0` structure of type DS

Output

`sumL_pen` scalar, minimum of the empirical contrast function

```
proc(1)=infodiv(struct PV par0,struct DS d0);
local
a,C,datarray,dim,g,ga,gord,invg,L,m,M_T,n,order,p,part,phi,s,sig,Sigma_spec,
smallC,smallC0,smallC_smooth,sub,sub2,sumL,sumL_pen,theta,theta_array,theta_str,
theta_unpack;
datarray=d0.dataarray;
sig=pvUnpack(par0,"sigma");
```

```
order=getOrders(datarray);
p=datarray[4,2,1,1];
p=getMatrix(p,1|1|1);
dim=datarray[4,3,1,1];
dim=getMatrix(dim,1|1|1);
M_T=datarray[4,4,1,1];
M_T=getMatrix(M_T,1|1|1);
n=order[2];
g=datarray[1,.,1:p,1:p];
g=getArray(g,1);
sub=datarray[4,5,1,1:p];
smallC0=datarray[1,.,1,1];
smallC0=getArray(smallC0,1);
C=datarray[2,.,1,1:p];
C=getArray(C,1);
C=atranspose(C,1|3|2);
phi=datarray[3,.,1:dim,1];
m=getMatrix(datarray,4|1|1);
m=m[1:p];
Sigma_spec=datarray[5,1,1,1];
Sigma_spec=getMatrix(Sigma_spec,1|1|1);
theta=zeros(p,dim);
theta_unpack=zeros(p,dim);
a=arrayinit(n|p|1,1);
s=0;
do while s<p;
  s=s+1;
  if m[s]==0;
    goto jump2;
  endif;
  theta_unpack[s,1:m[s]]=pvUnpack(par0,theta_str[s]);
  jump2:
endo;
theta_array=areshape(theta_unpack,n|p|dim);
phi=getArray(phi,1);
a=amult(theta_array,phi);
ga=amult(g,a);
gord=getorders(g);
invg=g*0;
for i(1,gord[1],1);
  part=invswp(getmatrix(g,i));
  invg[i,.,.]=part;
endfor;
L=ln(sig^2)+(1/sig^2)*(amult(atranspose((ga+C),1|3|2),amult(invg,(ga+C)))
  +smallC0-amult(atranspose(C,1|3|2),amult(invg,C)));
L=atranspose(L,2|1|3);
L=getMatrix(L,1);
sumL=1/(2*n)*sumc(L);
sumL_pen=sumL+pen(m, M_T, p, n, Sigma_spec);
retp(sumL_pen);
endp;
```

B.1.4 pen()

```
proc(1)=pen(m,M_T,p,n,Sigma_spec);
local c_3,c_4,d_m,d_m_type,L_m,pen_m,pen_m_test;
c_3=1;
c_4=1;
L_m=1;
d_m=sumc(m');
pen_m=c_3*d_m/n+c_4*d_m*(1+L_m)/n*Sigma_spec^2;
retp(pen_m);
endp;
```

B.1.5 count()

```
proc(1)=count(pmat,dim,sub);
local i,p;
p=rows(pmat);
if pmat[p]<dim and sub[p]==1;
  pmat[p]=pmat[p]+1;
else;
  pmat[p]=0;
  for i (p-1,1,-1);
    if pmat[i+1]==0;
      if pmat[i]<dim and sub[i]==1;
        pmat[i]=pmat[i]+1;
        goto jump;
      else;
        pmat[i]=0;
      endif;
    endif;
  endfor;
endif;
jump:
retp(pmat);
endp;
```

B.1.6 count_uneq_zero()

```
proc(1)=count_uneq_zero(x);
local i,n;
n=0;
i=0;
do while i<rows(x);
  i=i+1;
  if x[i]/=0;
    n=n+1;
  endif;
endo;
retp(n);
endp;
```

B.2 Procedures for computing the coefficient functions

B.2.1 coeff_theta()

```
proc(1)=coeff_theta(u,theta,min_m);
local a,j;
j=seqa(0,1,maxc(min_m'));
a =theta[1:cols(min_m),1:maxc(min_m')]*(sqrt(2)*cos(pi*j*u'));
retp(a');
endp;
```

B.2.2 coef()

```
proc(1)=coef(u,theta,min_m);
local a;
a=-1.8*cos(1.5-cos(2*pi*u));
a=a~(0.81*ones(rows(a),1));
retp(a);
endp;
```

B.2.3 coef_leg()

```
proc(1)=coef_leg(u,theta,min_m);
local a,p;
p=zeros(rows(u),cols(theta));
if cols(theta)>0;
  p[.,1]=ones(rows(u),1);
endif;
if cols(theta)>1;
  p[.,2]=u;
endif;
if cols(theta)>2;
  p[.,3]=1/2*(3*u^2-1);
endif;
if cols(theta)>3;
  p[.,4]=1/2*(5*u^3-3*u);
endif;
if cols(theta)>4;
  p[.,5]=1/8*(35*u^4-30*u^2+3);
endif;
if cols(theta)>5;
  p[.,6]=1/8*(63*u^5-70*u^3+15*u);
endif;
if cols(theta)>6;
  p[.,7]=1/16*(231*u^6-315*u^4+105*u^2-5);
endif;
if cols(theta)>7;
  p[.,8]=1/16*(429*u^7-693*u^5+315*u^3-35*u);
endif;
if cols(theta)>8;
```

```
  p[.,9]=1/128*(6435*u^8-12012*u^6+6930*u^4-1260*u^2+35);
endif;
if cols(theta)>9;
  p[.,10]=1/128*(12155*u^9-25740*u^7+18018*u^5-4620*u^3+315*u);
endif;
if cols(theta)>10;
  p[.,11]=1/256*(46189*u^10-109395*u^8+90090*u^6-30030*u^4+3465*u^2-63);
endif;
if cols(theta)>11;
  "Please choose fewer columns!";
endif;
a=p*theta';
retp(a);
endp;
```

B.2.4 TVAR()

Purpose

Simulating a TVAR process

Input

`&c` coefficient function
`theta` (r, s)-matrix of r parameter vectors with maximal length s
`min_m` $(1, r)$-matrix, lengths of parameter vectors
`s` variance
`n` number of observations
`factor` scalar, factor determining the sample:
  if 0.5, $[0, 0.5]$;
  if 0.9, $[0, 1 - \texttt{pred}/\texttt{n}]$;
  if 1, $[0, 1]$

Output

$(n, 1)$-vector, realization of the TVAR process

```
proc(1)=TVAR(&c,theta,min_m,s,n,pred,factor);
local e,p,t,u,x;
local c:proc;
p=rows(theta);
e=rndn(n+p+100,1)*sqrt(s);
x=e;
t=p;
do while t<n+p+100;
  t=t+1;
  if factor==0.5;
    u=((t-p-100)/(2*n))*(n/(n-pred));
  elseif factor==0.9;
    u=(t-p-100)/n;
  elseif factor==1;
```

```
    u=((t-p-100)/n)*(n/(n-pred));
  endif;
  x[t]=-c(u,theta,min_m)*x[t-1:t-p]+e[t];
endo;
retp(x[p+100+1:p+100+n]);
endp;
```

B.2.5 acoef()

```
proc(1)=acoef(&c,theta,min_m,n);
local a,p,t,u;
local c:proc;
p=2;
a=zeros(n+p+100,rows(theta));
t=p;
do while t<n+p+100;
  t=t+1;
  u=(t-p-100)/n;
  a[t,.]=c(u,theta,min_m);
endo;
tsplot(a,"",1);
retp(a[p+100+1:p+100+n,.]);
endp;
```

B.3 Procedures for computing different measures of forecast accuracy

B.3.1 Mean error

```
proc(1)=me(y,y_hat);
local me,k;
k=rows(y);
me=(y-y_hat)';
retp(me);
endp;

proc(1)=me1(y,y_hat);
local me,k;
k=rows(y);
me=sumc(y-y_hat)/k;
retp(me);
endp;
```

B.3.2 Mean absolute deviation

```
proc(1)=mad(y,y_hat);
local mad;
mad=sumc(abs(y-y_hat))/(rows(y));
retp(mad);
endp;
```

B.3.3 Mean square error

```
proc(1)=mse(y,y_hat);
local mse,k;
k=rows(y);
mse=((y-y_hat)^2);
retp(mse);
endp;

proc(1)= mse1(y,y_hat);
local mse1;
mse1=sumc((y-y_hat)^2)/(rows(y));
retp(mse1);
endp;
```

B.4 Procedures simulating and applying the model selection procedure and predictions for TVAR processes

B.4.1 simultvar()

Purpose

Simulating the model selection procedure and predictions for TVAR processes

Input

`T` scalar, length of time series
`n` scalar, number of simulations
`dim` scalar, non-negative integer by which the dimension of the spaces is bounded from above
`m_max` scalar, maximum length of parameter vectors in the model selection procedure
`&c` proc, pointer to the procedure for computing the coefficients
`funct` if 1, `&c = &coeff_theta`;
 if 2, `&c = &coef`;
 if 3, `&c = &coef_leg`
`theta` (r,s)-matrix, r parameter vectors (of length s) for the simulated process
`min_m_exct` $(1,r)$-matrix, (real) lengths of parameter vectors
`s_exct` scalar, (real) variance
`pred` number of prediction steps
`pm` scalar, forecasting method
 if `pm = 1,` coefficients computed for each specific time point;
 if `pm = 2,` constant coefficients;
 if `pm = 3,` recursive computation of forecasts
`M_T` scalar, number of off-diagonals of the preestimated covariance matrix which is used in the computation of the penalty function
`print_cov` if 1, only the covariance matrix is computed and printed;
 if 0, complete simulation
`myoutput` string, path/name of the output file
`bandw` scalar, non-negative integer $\leq n$, bandwidth used for the smoothing step

Output

results are plotted to the output file "myoutput"

```
proc(0)=
  simultvar(T,n,dim,m_max,&c,funct,theta,min_m_exct,s_exct,pred,pm,M_T,
            print_cov,m_min_chosen,factor,myoutput,bandw,freq);
local c:proc;
local
acfmat,arraydim,au,au_exct,au_ges,au_me,au_me1,au_mean,au_mse,au_mse1,au_pred,
au_pred_ges,au_pred_mean,au_std,au2,au2_pred,au3,au3_pred,au4,au4_pred,
bandw_str,dim_str,error_ges,factor_str,freq_str,funct_str,i,j,jump1,jump2,k,l,m,
m_max_str,M_T_str,m1,m12,m13,m14,m2,m3,m4,max_x_freq,max_x_freq_str,me_pred,
me_pred_ges,me_pred_max,me_pred_mean,me_pred_med,me_pred_min,me_pred_quant05,
me_pred_quant25,me_pred_quant75,me_pred_quant95,me_pred_std,me_pred2,me_pred3,
me_pred4,min_m,min_m_dim,min_m_ges,min_m_ges_order,min_m_mean,min_m_med,
min_m_med_ges,min_m_n,min_m_n_matrix,min_m_std,min_m_std_ges,mse_pred_med,
mse_pred_min,mse_pred_quant05,mse_pred_quant25,mse_pred_quant75,
mse_pred_quant95,mse_pred_std,mse_pred2,mse_pred3,mse_pred4,n_comb,n_comb_max,
n_str,order,outp,pacfmat,parV,parV2,parV3,parV4,pfad,pfad2,pfaddat,pfadplt,
pm_str,pred_c,pred_str,pred2_c,pred3_c,pred4_c,q,r,r1,r2,r3,r4,s,sigma_ges,
sigma_mean,sigma_med,sigma_std,T_max,T_max_freq,T_max_freq_str,T_str,theta_ges,
theta_ges_4plus,theta_ges_dim,theta_ges_plus,theta_mean,theta_mean_ges,
theta_med,theta_med_ges,theta_std,theta_std_ges,thetadim,thetadim3,u_plot,
usqu_ges,usqu_n,usqu_n2,usqu_n3,usqu_n4,usqu_z,usqu_z2,usqu_z3,usqu_z4,x,x_pred,
x_pred_ahead,x_pred_ahead2,x_pred_ahead3,x_pred_ahead4,x_pred2,x_pred3,x_pred4,
x2,x3,x4,xges,xges2,xges3,xges4;
screen OFF;
output file=^myoutput reset;
"Program execution:";
datestr(date) " at " timestr(time);
sigma_ges=zeros(n,1);
arraydim=zeros(3,1);
arraydim[1]=dim;
arraydim[2]=n;
arraydim[3]=m_max;
theta_ges=arrayinit(arraydim,0);
min_m_ges=zeros(n,dim);
mse_ges=zeros(n,pred);
mse_pred_ges=zeros(n,pred);
me_pred_ges=zeros(n,pred);
usqu_ges=zeros(n,2);
error_ges=zeros(n,1);
au_ges=arrayinit(n|dim|T,0);
au_pred_ges=arrayinit(n|dim|pred,0);
au_exct=c(seqa(0,1,T)/T,theta,min_m_exct);
au_exct=areshape(au_exct,n|cols(min_m_exct)|T);
if print_cov==1;
  goto jump1;
endif;
jump1:
xges=TVAR(&c,theta,min_m_exct,s_exct,T+pred,pred,factor);
"xges_kompl"; xges;
x=xges[1:T];
```

```
{x_pred,parV,min_m,au,au_pred,acfmat,pacfmat,acfmat,pacfmat}=
  modelselect(x,dim,m_max,M_T,pred,pm,print_cov,m_min_chosen,bandw,freq);
au_pred_ges[i,.,.]=au_pred;
au_ges[i,.,.]=au';
x_pred_ahead=x_pred[(rows(x_pred)-(pred-1)):(rows(x_pred))];
"i";i;
"x_pred_ahead"; x_pred_ahead';
"xges"; xges[T+1:T+pred]';
mse_pred=mse(x_pred_ahead,xges[T+1:T+pred]);
me_pred=me(x_pred_ahead,xges[T+1:T+pred]);
mse_pred_ges[i,1:pred]=mse_pred;
me_pred_ges[i,1:pred]=me_pred;
error_ges[i]=er(x_pred_ahead,xges[T+1:T+pred]);
sigma_ges[i]=parV[1];
r1=0;
do while r1<cols(min_m);
  r1=r1+1;
  if min_m[r1]>0;
    m=min_m[r1];
    m1=sumr(min_m[1:r1])-m;
    theta_ges[r1,i,1:m]=parV[m1+2:1+m+m1];
  endif;
  min_m_ges[i,1:cols(min_m)]=min_m;
endo;
sigma_med=median(sigma_ges);
sigma_mean=meanc(sigma_ges);
sigma_std=stdc(sigma_ges);
order=1;
r=0;
do while r<cols(min_m_ges)-1;
  r=r+1;
  order=order|10^r;
endo;
min_m_ges_order=min_m_ges*order;
theta_ges_dim=getorders(theta_ges);
theta_ges_dim[3]=theta_ges_dim[3]+1;
theta_ges_plus=arrayinit(theta_ges_dim,0);
r=0;
do while r<theta_ges_dim[1];
  r=r+1;
  theta=getmatrix(theta_ges,r);
  theta=theta~min_m_ges_order;
  theta=sortc(theta,cols(theta));
  theta_ges_plus[r,.,.]=theta;
endo;
theta_ges_4plus=arrayinit(maxc(min_m_ges_order)|theta_ges_dim,0);
r=0;
do while r<maxc(min_m_ges_order);
r=r+1;
  q=0;
  do while q<rows(min_m_ges_order);
    q=q+1;
    if theta_ges_plus[1,q,cols(theta)]==r;
```

```
    theta_ges_4plus[r,.,q,.]=theta_ges_plus[.,q,.];
    endif;
  endo;
endo;
min_m_n=zeros(maxc(min_m_ges_order),1);
r=0;
do while r<maxc(min_m_ges_order);
  r=r+1;
  min_m_n_matrix=getmatrix(theta_ges_4plus,r|1);
  min_m_n[r]=count_uneq_zero(min_m_n_matrix[.,cols(theta)]);
endo;
"min_m_n";  min_m_n;
n_comb=count_uneq_zero(min_m_n);
theta_med=arrayinit(n_comb|m_max|dim,0);
theta_mean=arrayinit(n_comb|m_max|dim,0);
theta_std=arrayinit(n_comb|m_max|dim,0);
q=0;
do while q<n_comb;
  q=q+1;
  r=0;
  do while r<dim;
    r=r+1;
    theta_med[q,.,r]=median(getmatrix(theta_ges,r));
    theta_mean[q,.,r]=meanc(getmatrix(theta_ges,r));
    theta_std[q,.,r]=stdc(getmatrix(theta_ges,r));
  endo;
endo;
min_m_med=median(min_m_ges);
min_m_mean=meanc(min_m_ges);
min_m_std=stdc(min_m_ges);
mse_pred_min=minc(mse_pred_ges);
mse_pred_quant05=quantile(mse_pred_ges,0.05);
mse_pred_quant25=quantile(mse_pred_ges,0.25);
mse_pred_med=median(mse_pred_ges);
mse_pred_quant75=quantile(mse_pred_ges,0.75);
mse_pred_quant95=quantile(mse_pred_ges,0.95);
mse_pred_max=maxc(mse_pred_ges);
mse_pred_mean=meanc(mse_pred_ges);
mse_pred_std=stdc(mse_pred_ges);
me_pred_min=minc(me_pred_ges);
me_pred_quant05=quantile(me_pred_ges,0.05);
me_pred_quant25=quantile(me_pred_ges,0.25);
me_pred_med=median(me_pred_ges);
me_pred_quant75=quantile(me_pred_ges,0.75);
me_pred_quant95=quantile(me_pred_ges,0.95);
me_pred_max=maxc(me_pred_ges);
me_pred_mean=meanc(me_pred_ges);
me_pred_std=stdc(me_pred_ges);
au_mean=amean(au_ges,3);
au_std=astd(au_ges,3);
au_pred_mean=amean(au_pred_ges,3);
"sigma_med";            sigma_med';
"sigma_mean";           sigma_mean';
```

```
"sigma_std";              sigma_std';
"theta_ges";              theta_ges;
"theta_med";              theta_med;
"theta_mean";             theta_mean;
"theta_std";              theta_std;
"min_m_ges_order";        min_m_ges_order;
"min_m_ges";              min_m_ges;
"min_m_med";              min_m_med;
"min_m_mean";             min_m_mean;
"min_m_std";              min_m_std;
"mse_pred_ges";           mse_pred_ges;
"mse_pred_min";           mse_pred_min';
"mse_pred_quant05";       mse_pred_quant05;
"mse_pred_quant25";       mse_pred_quant25;
"mse_pred_med";           mse_pred_med';
"mse_pred_quant75";       mse_pred_quant75;
"mse_pred_quant95";       mse_pred_quant95;
"mse_pred_max";           mse_pred_max';
"mse_pred_mean";          mse_pred_mean';
"mse_pred_std";           mse_pred_std';
"me_pred_min";            me_pred_min';
"me_pred_quant05";        me_pred_quant05;
"me_pred_quant25";        me_pred_quant25;
"me_pred_med";            me_pred_med';
"me_pred_quant75";        me_pred_quant75;
"me_pred_quant95";        me_pred_quant95;
"me_pred_max";            me_pred_max';
"me_pred_mean";           me_pred_mean';
"me_pred_std";            me_pred_std';
"au_ges";                 au_ges;
"au_mean";                au_mean;
"au_std";                 au_std;
"au_pred_ges";            au_pred_ges;
"au_pred_mean";           au_pred_mean;
"au_exct";                au_exct[1,.,.];
"au_mse";                 diag(getmatrix(amean(amult(au_ges-
                            au_exct,atranspose(au_ges-au_exct,1|3|2)),3),1)/T)';
"au_me";                  amean(amean(abs(au_ges-au_exct),1),3);
T_str=ftos(T,"%*.*lf",1,0);
n_str=ftos(n,"%*.*lf",1,0);
dim_str=ftos(dim,"%*.*lf",1,0);
m_max_str=ftos(m_max,"%*.*lf",1,0);
pred_str=ftos(pred,"%*.*lf",1,0);
funct_str=ftos(funct,"%*.*lf",1,0);
pm_str=ftos(pm,"%*.*lf",1,0);
M_T_str=ftos(M_T,"%*.*lf",1,0);
factor_str=ftos(factor,"%*.*lf",1,0);
bandw_str=ftos(bandw,"%*.*lf",1,0);
freq_str=ftos(freq,"%*.*lf",1,5);
max_x_freq=1*freq;
max_x_freq_str=ftos(max_x_freq,"%*.*lf",1,5);
pfad="a1ges_"$+T_str$+n_str$+dim_str$+m_max_str$+pred_str$+funct_str$+pm_str$+M_
        T_str$+factor_str$+bandw_str$+freq_str;
```

```
pfad2="a1ges_"$+T_str$+n_str$+dim_str$+m_max_str$+pred_str$+funct_str$+pm_str$+M
        _T_str$+factor_str$+bandw_str$+freq_str;
pfadplt=pfad$+".plt";
output file=^pfadplt reset;
print "set size 0.5,0.5";
"unset key";
"set xrange [0:"$+max_x_freq_str$+"]";
"set xlabel '$u$'";
"set xtics out nomirror";
"set ytics rotate nomirror out offset 1,0";
if funct==3;
"plot 0.2-0.1*(x/"$+freq_str$+")+0.1*0.5*(3*(x/"$+freq_str$+")**
  2-1)+0.2*0.5*(5*(x/"$+freq_str$+")**3-3*x/"$+freq_str$+")
  +0.3*0.125*(35*(x/"$+freq_str$+")**4-30*(x/"$+freq_str$+")**2+3) w l lw 1 lt 1
  notitle, '"$+pfad2$+".dat' u 1:2 notitle with lines lw 1 lt 2";
elseif funct==2;
"plot -1.8*cos(1.5-cos(2*pi*(x/"$+freq_str$+"))) with lines lw 1 lt 1
  notitle, '"$+pfad2$+".dat' u 1:2 notitle with lines lw 1 lt 2";
elseif funct==1;
"plot sqrt(2)*0.2*1-0.1*sqrt(2)*cos(pi*1*(x/"$+freq_str$+"))+0.1*sqrt(2)*
  cos(pi*2*(x/"$+freq_str$+"))+0.2*sqrt(2)*cos(pi*3*(x/"$+freq_str$+"))
  +0.3*sqrt(2)* cos(pi*4*(x/"$+freq_str$+")) with lines lw 1 lt 1 notitle,
  '"$+pfad2$+".dat'
u 1:2 notitle with lines lw 1 lt 2";
endif;
"set terminal epslatex blacktext";
outp="set output '"$+pfad2$+".tex'";
outp;
"replot";
"unset output";
"set term pop";
pfaddat=pfad$+".dat";
output file = ^pfaddat RESET;
u_plot=freq*seqa(1,1,T)/(T+pred);
u_plot~getmatrix(au_mean,1|1)';
pfad="a2ges_"$+T_str$+n_str$+dim_str$+m_max_str$+pred_str$+funct_str$+pm_str$+M_
  T_str$+factor_str$+bandw_str$+freq_str;
pfad2="a2ges_"$+T_str$+n_str$+dim_str$+m_max_str$+pred_str$+funct_str$+pm_str$+M
  _T_str$+factor_str$+bandw_str$+freq_str;
pfadplt=pfad$+".plt";
output file=^pfadplt RESET;
print "set size 0.5,0.5";
"unset key";
"set xrange [0:1]";
"set xlabel '$u$'";
"set xtics out nomirror";
"set ytics rotate nomirror out offset 1,0";
if funct==3;
"plot[0:"$+max_x_freq_str$+"][0:0.5] 0.3 w l lw 1 lt 1 notitle,
  '"$+pfad2$+".dat' u 1:2 notitle with lines lw 1 lt 2";
elseif funct==2;
"plot [0:"$+max_x_freq_str$+"][0.5:1] 0.81 with lines lw 1 lt 1 notitle,
  '"$+pfad2$+".dat' u 1:2 notitle with lines lw 1 lt 2";
```

```
elseif funct==1;
"plot [0:"$+max_x_freq_str$+"][0:0.5] sqrt(2)*0.3 with lines lw 1 lt 1 notitle,
  '"$+pfad2$+".dat' u 1:2 notitle with lines lw 1 lt 2";
endif;
"set terminal epslatex blacktext";
outp="set output '"$+pfad2$+".tex'";
outp;
"replot";
"unset output";
"set term pop";
pfaddat=pfad$+".dat";
output file=^pfaddat RESET;
u_plot=freq*seqa(1,1,T)/(T+pred);
u_plot~getmatrix(au_mean,1|2)';
output file=^myoutput ON;
"Programmende:";
datestr(date) " um " timestr(time) " Uhr ";
retp();
endp;
```

B.4.2 practevaltvar()

```
proc(0)=
  practevaltvar(dataset,tbegin,T,sub,m_max,pred,pm,M_T,print_cov,
  zeta_min,zeta_max,roll,myoutput);
local
alpha,alpha1,arresid,arresid1,au_ges,au_pred,auges,aupred,b,covsubar,dif,
estcoeff,fitted,i,j,lagset,l1,min_m,mse_pred,mse_pred_ar,mse_pred_ar_vec,
mse_pred_naive,mse_pred_naive_vec,mse_pred_vec,myoutputges,n,parV,res,resid,s,x,
x_fitted,x_ges,x_pred,x_pred_ahead,x_pred_ahead_ar,x_pred_ahead_naive,x_pred_ar,
x_pred_naive,x_pred_naive_star,x_resid,x1,z,zeta,zeta_str;
myoutputges=myoutput$+".aus";
output file=^myoutputges reset;
load x[]=^dataset;
x=x[tbegin:tbegin-1+T];
outwidth 256;
format /rd 12,8;
screen off;
x=x+0.0001*ones(rows(x),1);
zeta=zeta_min-0.1;
do while zeta<zeta_max-0.0001;
  resid=zeros(rows(x)-pred-roll,roll);
  fitted=zeros(rows(x)-pred-roll,roll);
  estcoeff=zeros(sumc(sub*m_max),roll);
  mse_pred=zeros(pred,roll);
  mse_pred_naive=zeros(pred,roll);
  mse_pred_ar=zeros(pred,roll);
  auges=seqa(1,1,T-roll+1-pred);
  aupred=seqa(1,1,pred);
  zeta=zeta+0.1;
  s=-1;
  do while s<roll-1;
```

```
    s=s+1;
    zeta_str=ftos(zeta*10,"%*.*lf",1,0);
          x_ges=x[s+1:rows(x)-roll+1+s];
          x1=x_ges[1:rows(x_ges)-pred];
          n=rows(x1);
    z=zeros(1,m_max);
{x_pred,x_fitted,x_resid,parV,min_m,au_ges,au_pred}=
  modelselect(x1,sub,m_max,M_T,pred,pm,print_cov,z,rows(x1),zeta);
    auges=auges~au_ges;
    aupred=aupred~au_pred';
    {x_pred_naive_star,b}=exposmooth(x1,0,1,pred);
    x_pred_naive=x_pred_naive_star[.,2];
    arimaset;
    {lagset,alpha,covsubar,res}=
      arsubfit(x1[rows(x1)-49:rows(x1)],indexcat(sub,1),0,rows(sub));
    alpha1=zeros(maxc(lagset),1);
    alpha1[lagset]=alpha;
    arresid=armaresid(x1[rows(x1)-49:rows(x1)],alpha1,0);
    arresid1=arresid[1:rows(sub)]|arresid;
    x_pred_ar=
      tsforecast(alpha1,x1[rows(x1)49:rows(x1)],rows(sub),0,0,0,arresid1,pred);
    x_pred_ahead=x_pred[(rows(x_pred)-(pred-1)):(rows(x_pred))];
    x_pred_ahead_naive=x_pred_naive[(rows(x_pred)-(pred-1)):(rows(x_pred))];
    x_pred_ahead_ar=x_pred_ar[.,2];
    "zeta"; zeta;
    dif=x_pred_ahead-x_ges[n+1:n+pred];
    mse_pred[.,s+1]=mse(x_pred_ahead,x_ges[n+1:n+pred]);
    mse_pred_naive[.,s+1]=mse(x_pred_ahead_naive,x_ges[n+1:n+pred]);
    mse_pred_ar[.,s+1]=mse(x_pred_ahead_ar,x_ges[n+1:n+pred]);
    "min_m"; min_m;
    "mse_pred"; mse_pred[.,s+1];
    resid[1:rows(x_resid),s+1]=x_resid;
    fitted[1:rows(x_fitted),s+1]=x_fitted;
    estcoeff[1:rows(parV')-1,s+1]=parV[2:rows(parV')]';
    mse_pred_vec=zeros(pred,1);
    i=0;
    do while i<pred;
            i=i+1;
      mse_pred_vec[i]=meanc(mse_pred[i,.]');
    endo;
    mse_pred_naive_vec=zeros(pred,1);
    j=0;
    do while j<pred;
            j=j+1;
      mse_pred_naive_vec[j]=meanc(mse_pred_naive[j,.]');
    endo;
    mse_pred_ar_vec=zeros(pred,1);
    j=0;
    do while j<pred;
            j=j+1;
      mse_pred_ar_vec[j]=meanc(mse_pred_ar[j,.]');
    endo;
  endo;
```

```
  "TVAR";
  format /ldn 12,0;
  seqa(1,1,pred)';; "  average  ";
  "MSE";
  format /ldn 12,8;
  mse_pred_vec';; meanc(mse_pred_vec);
  "naive";
  format /ldn 12,0;
  seqa(1,1,pred)';; "  average  ";
  "MSE";
  format /ldn 12,8;
  mse_pred_naive_vec';; meanc(mse_pred_naive_vec);
  "AR";
  format /ldn 12,0;
  seqa(1,1,pred)';; "  average  ";
  "MSE";
  format /ldn 12,8;
  mse_pred_ar_vec';; meanc(mse_pred_ar_vec);
  "mse_pred"; mse_pred;
  myoutputges=myoutput$+zeta_str$+"res.aus";
  output file=^myoutputges RESET;
  resid;
  myoutputges=myoutput$+zeta_str$+"fit.aus";
  output file=^myoutputges RESET;
  fitted;
  myoutputges=myoutput$+zeta_str$+"aug.aus";
  output file=^myoutputges RESET;
  auges;
  myoutputges=myoutput$+zeta_str$+"aup.aus";
  output file=^myoutputges RESET;
  aupred;
  myoutputges=myoutput$+zeta_str$+"cof.aus";
  output file=^myoutputges RESET;
  estcoeff;
  myoutputges=myoutput$+zeta_str$+"mse.aus";
  output file=^myoutputges RESET;
  mse_pred_vec';
  mse_pred_naive_vec';
  mse_pred_ar_vec';
  myoutputges=myoutput$+".aus";
  output file=^myoutputges ON;
endo;
retp();
endp;
```

Volkswirtschaftliche Analysen

Herausgegeben von Elisabeth Allgoewer, Georg Hasenkamp, Wolfgang Maennig, Christian Scheer und Peter Stahlecker

Band 1 Matthias Ross: Unsicherheit und Arbeitsnachfrage. Eine theoretische und empirische Analyse des warenproduzierenden Gewerbes in Deutschland. 1998.

Band 2 Olaf Schlotmann: Die deutsche Zeitstruktur der Zinssätze im Lichte der Wicksellschen Kredittheorie. 1998.

Band 3 Gunar Schröer: Schätz- und Testverfahren im strukturellen Kointegrationsmodell. 1998.

Band 4 Joachim Dannenbaum: Der Strukturwandel in Deutschland und seine Bedeutung für die Arbeitslosigkeit. 1998.

Band 5 Birgit Layes: Kompatibilität von Arbeitsqualifikationen. Eine volkswirtschaftliche Analyse. 1998.

Band 6 Nils Hauenschild: Alterssicherungssysteme im Rahmen stochastischer Modelle überlappender Generationen. 1999.

Band 7 Axel Kunde: Langfristige Auswirkungen der wirtschaftlichen Integration der Mitgliedsstaaten der Europäischen Union. 2000.

Band 8 Anna-Maria Herfeld: Reformansätze zur Alterssicherung in Deutschland und intergenerationelle Gerechtigkeit. 2001.

Band 9 Oliver W. Wojahn: Airline Networks. 2001.

Band 10 Thomas Krause: Dynamische Straßenbenutzungsgebühren zur Reduzierung von Staus. 2003.

Band 11 Markus Klintworth: Schätzverfahren im linearen Regressionsmodell bei partiellen und unscharfen Parameterrestriktionen. 2004.

Band 12 Harm Bandholz: Konjunkturelle Frühindikatoren für die Bundesrepublik Deutschland und die Freie und Hansestadt Hamburg. Eine empirische Analyse unter Verwendung linearer und nichtlinearer dynamischer Faktormodelle. 2004.

Band 13 Heike Ruhrmann: Nationale Innovationssysteme. Eine komparative Analyse am Beispiel Irlands und Finnlands. 2007.

Band 14 Elise Hefti: Economic Aspects of Obesity. 2009.

Band 15 Sonja Scheffler: Das Bild des Unternehmers im Spiegel der Gewerbepolitik im nachnapoleonischen Preußen. 2009.

Band 16 Jens-Ragnar Martinen: Die Tonnagesteuer. Eine ökonomische Analyse. 2010.

Band 17 Angaben folgen

Band 18 Cristina Sattarhoff: Statistical Inference in Multifractal Random Walk Models for Financial Time Series. 2011.

Band 19 Tina Loll: Forecasting Economic Time Series using Locally Stationary Processes. A New Approach with Applications. 2012.

www.peterlang.de